Chemistry of Food, Food Supplements, and Food Contact Materials: From Production to Plate

ACS SYMPOSIUM SERIES **1159**

Chemistry of Food, Food Supplements, and Food Contact Materials: From Production to Plate

Mark A. Benvenuto, Editor
University of Detroit Mercy, Detroit, Michigan

Satinder Ahuja, Editor
Ahuja Consulting, Calabash, North Carolina

Timothy V. Duncan, Editor
US Food and Drug Administration, Bedford Park, Illinois

Gregory O. Noonan, Editor
US Food and Drug Administration, College Park, Maryland

Elizabeth S. Roberts-Kirchhoff, Editor
University of Detroit Mercy, Detroit, Michigan

Sponsored by the
ACS Division of Environmental Chemistry
ACS Division of Agricultural and Food Chemistry

American Chemical Society, Washington, DC

Distributed in print by Oxford University Press

Library of Congress Cataloging-in-Publication Data

Chemistry of food, food supplements, and food contact materials : from production to plate / Mark A. Benvenuto, editor, University of Detroit Mercy, Detroit, Michigan, Satinder Ahuja, editor, Ahuja Consulting, Calabash, North Carolina, Timothy V. Duncan, editor, US Food and Drug Administration, Bedford Park, Illinois, Gregory O. Noonan, editor, US Food and Drug Administration, College Park, Maryland, Elizabeth S. Roberts-Kirchhoff, editor, University of Detroit Mercy, Detroit, Michigan ; sponsored by the ACS Division of Environmental Chemistry, ACS Division of Agricultural and Food Chemistry.
 pages cm. -- (ACS symposium series ; 1159)
 Includes bibliographical references and index.
 ISBN 978-0-8412-2952-5
 1. Food--Analysis--Congresses. 2. Food--Composition--Congresses. I. Benvenuto, Mark A. (Mark Anthony), editor of compilation. II. American Chemical Society. Division of Environmental Chemistry. III. ACS Division of Agricultural and Food Chemistry.
 TX541.C437 2014
 664'.07--dc23

       2014014403

Foreword

The ACS Symposium Series was first published in 1974 to provide a mechanism for publishing symposia quickly in book form. The purpose of the series is to publish timely, comprehensive books developed from the ACS sponsored symposia based on current scientific research. Occasionally, books are developed from symposia sponsored by other organizations when the topic is of keen interest to the chemistry audience.

Before agreeing to publish a book, the proposed table of contents is reviewed for appropriate and comprehensive coverage and for interest to the audience. Some papers may be excluded to better focus the book; others may be added to provide comprehensiveness. When appropriate, overview or introductory chapters are added. Drafts of chapters are peer-reviewed prior to final acceptance or rejection, and manuscripts are prepared in camera-ready format.

As a rule, only original research papers and original review papers are included in the volumes. Verbatim reproductions of previous published papers are not accepted.

ACS Books Department

Contents

1. Overview 1
   M. A. Benvenuto, S. Ahuja, T. Duncan, G. O. Noonan, and E. Roberts-Kirchhoff

2. Sustainability in Foods and Food Production: The Roles of Peer Reviewed
   Science and Corporate Initiatives 5
   James N. Seiber and Loreen Kleinschmidt

3. The New Inconvenient Truth: Global Contamination of Food by Chemical
   Pollutants, Particularly Heavy Metals and Metalloids 15
   Tsanangurayi Tongesayi and Sunungurai Tongesayi

4. FDA's Regulation of Nanotechnology in Food Ingredients 41
   Teresa A. Croce

5. A Comprehensive Study into the Migration Potential of Nano Silver
   Particles from Food Contact Polyolefins 51
   J. Bott, A. Störmer, and R. Franz

6. Food Packaging: Strategies for Rapid Phthalate Screening in Real Time by
   Ambient Ionization Tandem Mass Spectrometry 71
   Elizabeth Crawford, Catharina Crone, Julie Horner, and Brian Musselman

7. Examination of a Selection of the Patent Medicines and Nostrums at
   the Henry Ford Museum via Energy Dispersive X-ray Fluorescence
   Spectrometry 87
   Andrew Diefenbach, Danielle Garshott, Elizabeth MacDonald, Thomas Sanday,
   Shelby Maurice, Mary Fahey, and Mark A. Benvenuto

8. Analysis of Nine Edible Clay Supplements with a Handheld XRF Analyzer 99
   Jessica L. LaBond, Nicholas H. Stroeters, Mark A. Benvenuto, and
   Elizabeth S. Roberts-Kirchhoff

Editors' Biographies 113

Indexes

Author Index 117

Subject Index 119

Chapter 1

Overview

M. A. Benvenuto,[1] S. Ahuja,[2] T. Duncan,[3] G. O. Noonan,*,[4] and E. Roberts-Kirchhoff[1]

[1]University of Detroit Mercy, Chemistry & Biochemistry, 4001 W. McNichols Rd., Detroit, Michigan 48221
[2]1061 Rutledge Court, Calabash, North Carolina 28467
[3]US Food & Drug Administration, 6502 S. Archer Road, Bedford Park, Illinois 60516
[4]US Food & Drug Administration, 5100 Paint Branch Parkway, College Park, Maryland 20740
*E-mail: Gregory.Noonan@fda.hhs.gov

Ensuring the safety of food requires a complex and ever-changing set of interactions between producers, distrbutors, consumers and regulators. As advances are made in packaging and food additives, as food distributions systems evolve to meet consumer needs, or as these respond to environmental and population changes, adjustments to regulatory systems may become neceesary. Analytical, environmental and materials chemistry can often play important roles in responding to these changes and in continuing to help with the improvement of food safety and security. This chapter gives a breif overview of the chapters, produced from seminars presented at three symposia held at the 245th ACS National Convention in New Orleans, Louisiana in the spring of 2013.

For more than a century, national and international governing bodies have had some involvement in regulating the quality and safety of food during production and delivery. In the United States, a common belief is that Upton Sinclair's seminal work, "The Jungle," was the driving force behind quality control and manufacturing standards within the entire food production industry and spurred Congress to pass the Pure Food and Drug Act in 1906 (*1, 2*). While Sinclair's influence may have been more public, early pioneers in food safety, including Peter Collier and Harvey Wiley in the US and Friedrich Accum in Germany and the UK (3), had been working toward ensuring the safety and quality of food for decades before Sinclair's novel was published. Whatever the catalysts for improved oversight, it is clear that "modern" food regulation saw its start at the beginning of the 20th century. Since then, the way food is produced, packaged and distributed has undergone dramatic changes and the rate of change has progressively increased. For instance in the last 30 years there has been tremendous growth in new food products containing new additives and often presented in new packaging materials. Additionally, the food distribution system is no longer described as a food chain, but is more often referred to as a "food web" and includes a complex international structure with production and packaging often happening at multiple points and in multiple countries between the farm and table. Along with all of the changes to food production and distribution there have been changes to national and international controls and regulations. For example the passage of the Food Safety and Modernization Act in the US (4) and Plastics Regulation in the EU (5) are recent examples of numerous updates to the national/regional regulatory overview. On an international level, the Codex Alimentarius Commission, whose goal is to *"develop harmonised international food standards, guidelines and codes of practice to protect the health of the consumers and ensure fair practices in the food trade"* (6) just celebrated its 50th anniversary. Clearly changes in food production and distribution have been met by changes in the regulatory environment, however it is fair to ask that as the pace of change within the food industry becomes more rapid, due to technological advances or environmental changes, will regulation be able to continue to address emerging food quality and food safety issues.

It is difficult to determine if technological advances in the areas of polymer science, refrigeration, and transportation have driven the globalization of the food supply or if the food industry has drawn from these technologies to satisfy consumer's desire and need. Whatever the driving force, it is clear that "food miles" have increased dramatically in the past few decades and that technology has enabled this increase. Food packaging, often referred to as food contact materials (FCM) is one area that has benefitted from technological advances. What started thousands of years ago as simple items such as gourdes, animal hides, and baskets have evolved into increasingly complex multicomponent/multilayer materials (7, 8) that enable the long term storage of foods without the need for refrigeration. One consequence of food packaging, is that components from the FCM often migrate into the food during processing (filling, sterilizing) and storage. This migration is often surprising and disturbing to consumers (9), however the scientific and regulatory community has recognized, studied and regulated the migration of compounds from FCMs for decades (10). As new, advanced FCMs are developed, there is a need to assess if traditional models/theories used to assess the safety of the products are applicable to the new materials. Currently, the incorporation of nanomaterials into food packaging is presenting a new challenge to scientists and regulators. While nanomaterials clearly require new analytical characterization tools (11), there is still some disagreement if traditional migration models can be used to predict the migration of nanomaterials from polymer substrates. Within this volume, Croce discusses the position and response of the US FDA to the safety and use of nanomaterials as food additives, either directly or as migrants from FCMs. Additionally, Bott, Stoermer and Franz systematically evaluate the potential migration of nanosilver from FCMs and address if traditional migration models are relevant to these new materials.

One of the side effects of the increased food miles is the dramatic increase in the quantity of food crossing borders, often moving from countries with low and poorly regulated production facilities to areas of high and well regulated systems. This increase could quickly tax existing inspection and testing regimes. However to offset the increase, there has been a response to update analytical methods and implement technology that reduces per-sample testing times and increases sample throughput. Direct analysis in real time mass spectrometry (DART-MS) (*12*) and X-ray fluorescence (XRF) (*13*) are two relatively new techniques that require limited sample preparation and have potential for increasing sample throughput. Crawford, Crone, Horner, Musselman describe the use of DART-MS to the detection of phthalates, compounds commonly used in FCMs. The utility of XRF is demonstrated in 2 chapters, by applying XRF for the determination of metal concentrations of clay supplements and historical patent medicines.

The increased food distribution system brings benefits, such as access to native foods and to a wide variety of fresh fruits and vegetables throughout the year. However, it also raises questions about economic stability of developing nations, environmental contamination and sustainability. These are complex issues that include political, economic and cultural aspects entangled with food safety. Two chapters consider globalization, with Tongesayi and Tongesayi using toxic element contamination to address economic, political and natural resource issues and the role they play in food safety. Seiber and Kleinschmidt offer a wider view, discussing how science, corporate initiatives and consumer views of food safety and sustainability are intertwined.

The discussion on food, food safety, dietary supplements, and the chemistry involved in all these processes and situations, has been long, wide-ranging and ongoing (*14–16*). Changes in production, distribution and waste (recycling) can lead to changes in consumer perceptions and concerns (*17, 18*). In this volume, produced from seminars presented at three symposia held at the 245th ACS National Convention in New Orleans, Louisiana in the spring of 2013, we have papers on the ability to screen for phthalate additives, the FDA's regulation of nano-technology incorporated into foods, trace materials that can be found in edible clays, how silver nano-particles migrate in food packaging, and patent medicines that have been stored for over a century. By drawing on diverse areas connected to food production and distribution we hope that this volume will be useful in the continuing discussion concerning food safety and preservation, and in highlighting the role chemistry can play in ensuring that all individuals have access to safe food.

References

1.   Sinclair, U. *The Jungle*; Doubleday: 1906.

2. Pure Food and Drug Act 21 § U.S.C, 1906.

3. Accum, F. C. *A treatise on adulterations of food and culinary poisons*; J. Mallet: London, 1820.

4. FDA Food Safety Modernization Act, 2011

5. European Commission. Commision Regulation 10/2011 of 14 January 2011 on to plastic materials and articles intended to come into contact with food. *Off. J. Eur. Communities* **2011**, *12*, 1–89.

6. http://www.codexalimentarius.org/ (accessed on March 28, 2014).

7. Risch, S. J. Food Packaging History and Innovations. *J. Agric. Food Chem.* **2009**, *57* (18), 8089–8092.

8. Marsh, K.; Bugusu, B. Food Packaging—Roles, Materials, and Environmental Issues. *J. Food Sci.* **2007**, *72* (3), R39–R55.

9. Boggan, S. Poisoned by plastic: Chemicals in water bottles and food packaging have been June linked to infertility and birth defects. Scaremongering, or the truth? *Mail Online*, http://www.dailymail.co.uk/ health article-2157423/ (accessed March 28, 2014).

10. Federal Food, Drug, and Cosmetic (FD&C) Act.

11. Dudkiewicz, A.; Tiede, K.; Loeschner, K.; Helene, L.; Jensen, S.; Jensen, E.; Wierzbicki, R.; Boxall, A. B. A.; Molhave, K. Characterization of nanomaterials in food by electron microscopy. *Trends Anal. Chem.* **2011**, *30* (1), 28–43.

12. Cody, R. B.; Laramee, J. A.; Durst, H. D. Versatile new ion source for the analysis of materials in open air under ambient conditions. *Anal. Chem.* **2005**, *77* (8), 2297–2302.

13. Palmer, P. T.; Jacobs, R; Baker, P. E.; Ferguson, K.; Webber, S. Use of Field-Portable XRF Analyzers for Rapid Screening of Toxic Elements in FDA-Regulated Products. *J. Agric. Food Chem.* **2009**, *57* (7), 2605–2613.

14. Jackson, L. S. Chemical Food Safety Issues in the United States: Past, Present, and Future. *J. Agric. Food Chem.* **2009**, *57* (18), 8161–8170.

15. Sen, N. P. Migration and Formation of *N*-Nitrosamines from Food Contact Materials *Food and Packaging Interactions*; ACS Symposium Series; American Chemical Society: Washington, DC, 1988; Volume 365, Chapter 12, pp 146–158.

16. Muralidhara, H. S.; Satyavolu, J. Reducing Food-Processing Costs in the 21st Century: Need for Innovative Separation Technologies. *Ind. Eng. Chem. Res.* **1999**, *38* (10), 3710–3714.

17. Tacito, L. D. Polymer Recycling Technology for Food-Use Technical Requirements To Meet Safety and Quality Assurance. *Plastics, Rubber, and Paper Recycling*; ACS Symposium Series; American Chemical Society: Washington, DC, 1995; Volume 609; Chapter 39, pp 488–500.

18. Recycled Plastics for Food-Contact Applications: Science, Policy, and Regulation, Kuznesof, P. M., VanDerveer, M. C. *Plastics, Rubber, and Paper Recycling*; ACS Symposium Series; American Chemical Society: Washington, DC, 1995; Volume 609, Chapter 32, pp 389–403.

Chapter 2

Sustainability in Foods and Food Production: The Roles of Peer Reviewed Science and Corporate Initiatives

James N. Seiber* and Loreen Kleinschmidt

Department of Environmental Toxicology, University of California, Davis, Davis, California 95616
*E-mail: jnseiber@ucdavis.edu

There is much current interest in the term 'sustainability' in the production and use of foods and beverages. For food production and processing, sustainability can refer to foods that optimize health, safety, quality, and consumer appeal, as well as reduced inputs of energy, fertilizers, pesticides, and water during production or processing. It can include minimizing waste generation from the food itself or its packaging, reducing emissions, reducing the carbon footprint, humane treatment of farm animals, recycling waste for energy recovery, capturing and using wastewater and rainwater, and pest control with biopesticides rather than synthetics. A number of corporations, such as Walmart, have instituted goals relating to sustainability. This chapter summarizes some recent research into the benefits of implementing sustainable practices, including reduced-risk biopesticides or chemical-free pest control in sustainable systems of food production.

The term 'sustainability' is often used in connection with production and use of foods, as with many other products and activities of modern life. In food production, sustainability can refer to optimizing health, safety, quality and consumer appeal of sufficient foods to meet the demands of a growing base of consumers. The present world population of ca 7 billion has resulted largely from increases over the past 75 years, with the estimated rate of growth increasing slightly from about 1920-1950, then ramping up at a much faster rate since then (*1*). This accelerated pace of population growth, which will continue at least for the next 25 years, has been sustained by world food production increases of 20% from 1990 to 2010 (Figure 1) (*2*). This rate of increase of food production must continue in order for food sustainability to meet the population increases anticipated for at least the next 20 years. (Alternative scenarios are that the population of the world will level off at some time in the future, or that the world population will decline. The former seems most likely of these two scenarios, but it will take decades for this leveling off to occur. A decline in world population would occur if there is some drastic change, such as global climate change, world war, or volcanoes/earthquakes that catastrophically affect climate, water availability, or population—or all three.)

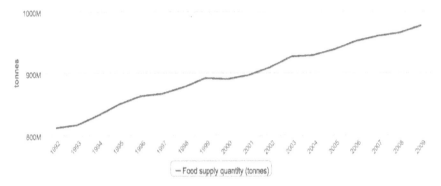

Figure 1. World food supply in tonnes, 1992-2009 (2). Source: Food and Agriculture Organization of the United Nations. FAOSTAT3.http:// faostat3.fao.org/faostatgateway/go/to/browse/C/*/E. Reproduced with permission.

The U.S. food and beverage industry is comprised of all companies involved in processing raw food materials, packaging, and distributing them. It is about a $1.5 trillion industry in the U.S (*3*). The market value of products sold for all farms in the U.S. has risen significantly, from $201 billion in 2002, rising to $297 billion in 2007, a 48% increase (*4*). The difference between $1.5 trillion and $297 billion reflects, approximately, the costs of processing, storing, and distributing foods at the wholesale and retail levels. Clearly we are presently witnessing significant growth in incomes in the farming sector, at least in the U.S.

Challenges faced by the industry have also risen, including costs associated with labor, water, energy, and the cost of compliance to increasing regulation. There is also increased competition in export markets. The industry is attempting to adapt to consumer interests, in such areas as the demand for green and locally-produced foods. So it is not just size, but many factors that define sustainability.

How to define and measure sustainability, and how food producers and distributors can best respond to the call for 'sustainability' are not easy questions to address. It depends on how sustainability is defined and on one's goals in food production and use. For researchers, particularly in the health and nutrition fields,

6

the health benefits of foods to the consumer are especially important (e.g., vitamin, antioxidant, minerals balance and avoiding undue levels of fats, carbohydrates, sodium and synthetic additives (5–11). Producers, processors, corporations, and consumers often have differing viewpoints on sustainability, with some overlap and some novel aspects, as mentioned in the sections which follow.

View from Producers

For producers, sustainability includes judicious management choices regarding land, water, and energy inputs, as well as fertilization and pest management tools. EPA registration of new pesticides is now largely devoted to biorational or biopesticides, in addition to modification to improve synthetics developed largely in the 20th century which have proved to be of major significance in increasing productivity at the farm level (12–14).

Examples from the websites of agricultural producers underscore other elements this group regards as important for sustainability. Far Niente Winery in the Napa Valley, California, bills itself and its sister wineries as following "an integrated program of sustainable measures affecting vineyard, winery and day-to-day business practices (15)." This includes solar powered "net-zero" use of electricity , organically farmed vineyards, reuse of process waters from winery operations for irrigation and frost protection, use of biodiesel fueled farm vehicles and hybrid engine vehicles, and extensive recycling efforts.

Dixon Ridge Farms in Winters, California, winner of the 2012 U.S. EPA Sustainable Agricultural Champion Award, as well as awards from California agencies, agricultural- and environmental groups, uses sustainable practices for both growing and processing (16, 17). This 1,400 acre organic walnut orchard employs low or no till and low mow practices to reduce energy use, retain moisture, and reduce runoff water loss. They recycle water and use soil moisture meters to direct irrigation rates. They recycle heat used in their drying operations, and gasify walnut shells to provide fuel for electric generators, as well as using solar panels. In addition to IPM techniques during the growing of the walnuts, they use an energy-efficient freezer coupled with a proprietary technique to kill insects, worms and eggs in the stored product. They also maintain a perennial insectary habitat adjacent to their orchards. They fertilize with composted manure, chipped tree prunings and walnut hulls, and use walnut shells as mulch, and maintain a vibrant cover crop on the orchard floor. They actively partner with many groups in teaching and research on sustainable practices for working farms.

View from Processors

For food processors, qualities valued include health, safety, and product quality, among many others such as shelf life, transportability, and, of course, taste and consumer appeal. Food processors need also be aware that some processing steps can lead to formation of Maillard reaction products in foods, many of which are beneficial in terms of taste and odor but a few of which are toxic. High temperature cooking of potatoes and cooking or roasting of other high carbohydrate foods can lead to formation of acrylamide (*18, 19*), diacetyl (*20*), and 4-methyl imidazole (*21*). Recent research shows ways to reduce acrylamide formation by a variety of approaches (*22, 23*). Pulegone is a natural constituent of mint which gets expressed in some mint-flavoring agents subsequently added to many products (*24*).

For both the producer and processor, sustainability may include more efficient use of irrigation water, for example by drip irrigation or hydroponics culture, or recovery of water from processing or sewage wastewaters. Sustainability may also include farm and processing machinery that uses biofuels or solar or wind power, integration of farming crops (e. g., rice) with salmon or duck production, and farm animal production with more humane treatment of poultry, beef, cows, and sheep. Advances in farm animal production efficiency in recent years in the U.S., whether in milk and other dairy products or in meat and poultry products, has been impressive (*25*). And it comes at a time when global consumption of meat and meat products is accelerating (e.g., KFC growth in China—a new franchise outlet opens on average every day!)

For food processors, sustainability is more about preserving and distributing foods, including maximizing shelf life. In just the area of transport and distribution of foods, U.S. food companies, including processors, have established amazingly quick turn-around in both distribution and traceability networks, using truck, train, and ship transports (*26*). Processors have also advanced significantly in the areas of quality, selection, flavor, aroma, convenience, and other market driven qualities. Since over 70% of U.S. foods are processed, processing goals have continued, even as the fresh and local movement gains momentum, perhaps because food processing plays a major role in the health, safety, quality, and consumer appeal of foods, in addition to availability and low cost.

Safety Aspects

The importance of 'safety' to sustainability pervades all aspects of the food supply. Safety cannot be overemphasized as shown in the loss of markets and consumer confidence following outbreaks of pathogen-related food poisonings, such as *E. coli* 0157H:7, *Salmonella*, and *Listeria* in fresh salad items, and processed meat and cheeses (*27*). Food safety testing and oversight continues to spur development of genotyping, near infrared and mass spectrometric methods, and other techniques for detecting and analyzing pathogens, acrylamide, diacetyl, as well as toxic elements like arsenic and mercury and other chemicals associated with mineral extraction, packaging, and preservation operations.

A current safety challenge is how to deal with low level toxic contaminants in foods that are due to natural or largely unavoidable scenarios. Mercury and other heavy metals, acrylamide and related Maillard browning reaction by-products are examples. Recent findings of arsenic levels in rice and other staple foods frame the issue. Arsenic in soil and water, albeit at very low levels, can contaminate grains like rice by plant uptake with nutrients (*28*). Arsenic exists in both organic and inorganic forms, the latter posing special challenges because of its proven carcinogenicity such that a safe level of exposure cannot be determined.

Consequently, the U.S. Food and Drug Administration (FDA) does not presently have an adequate scientific basis to recommend changes in consumption patterns of rice and rice products. Average levels of inorganic arsenic for rice and rice products surveyed recently by FDA range from 3.5 to 6.7 micrograms of inorganic arsenic per serving in the U.S (29). Rice from other nations, such as Bangladesh, with arsenic contaminated irrigation water in shallow ground table reserves, can exceed the levels found in U.S. rice. Consumer exposure from drinking water as well as foods like rice and other cereal grains, and also from air, may result in aggregate exposures that exceed the recommended daily intake of inorganic arsenic. How to deal with risks from low-level arsenic exposure is not yet clear. Consumer advisories can provide information, and processors can search for supplies of rice that minimize exposures in foods. There is interest in reducing arsenic levels in contaminated foodstuffs, like rice, but no convenient way of reducing arsenic in foods and beverages in general has been forthcoming. Recent research on application of silica-based foliage fertilizers to rice plants has shown promise as a way of reducing arsenic accumulation in rice grain (30).

Corporate Involvement

Food sustainability clearly includes safety, health, quality and efficiency elements, of both the foods and the environment in which it is produced. Sustainability must also preserve features of flavor, low cost, and wide availability that are so important to consumers. This raises an interesting question: Who is in the driver's seat on food sustainability? Presently the wholesaler/retailers are most visible, perhaps because of a perceived market advantage to having products with a 'sustainability' label that is aligned with consumer interests. Does it equate with the image of rampant consumption associated with a Walmart or Costco? There is evidence that sustainability is good business. A recent check on Walmart's latest sustainability index, which is experiencing growth in product categories, including food, and requirements of Walmart suppliers, shows the following goals:

- Increase in use of recycled materials to replace single use plastics and other items
- New products with greener chemicals, affecting products ranging from household cleansers, to personal care, beauty and cosmetic products
- Reducing fertilizer use in agriculture with a potential to reduce fertilizer use on 14 million acres in the U.S. by 2020
- Expanding the sustainability index to international markets currently in Chile, Mexico, and South Africa
- Improving Energy Efficiency, including in household appliances (31)

Walmart's environmental sustainability goals include: supply of 100% renewable energy; create zero waste; sell products that sustain people and the environment (32).

Consumers' View

For consumers, food sustainability often equates to local, organic, non-genetically modified (non-GMO) foods produced by family farms. One may question whether addressing these qualities can lead to the efficiency in production needed to ramp up food production to meet the needs of an anticipated 9 billion consumers worldwide by 2040 (*33*), without major increases in land, water, and other vital resources. GMO modifications, for example, can improve insect and disease resistance of crops as well as lessen the need for herbicides, all of which can improve yield and reduce use of pesticides.

Recent Research: Examples and Challenges

The peer reviewed literature increasingly contains articles on new biopesticides, biofuels, and biobased products used in the production, processing, and distribution of foods (*34*). Biopesticides promise a new generation of pest control agents that are of lower non-target toxicity, less residual, and produced often by fermentation rather that chemical synthesis relative to 'conventional' pesticides such as organochlorine and organophosphorus insecticides, phenoxy acid ester and triazine herbicides. The pest control targets become even more precise when transgenes inserted in crop plants are developed for pest management purposes (*35*).

Capturing and using the energy from processing and food wastes using anaerobic fermentation to biogas, and recovering the energy as electricity (with a by-product of compost and fertilizer) is an example of a 'green' bioenergy strategy (*36*). Another recent example is production of bioethanol and recovery of dietary supplements from grape pomace (*37, 38*), discovery of natural sources to control pests or as fertilizer (*39*) and use of natural materials as processing aids for clarification of beer or other beverages (*40*). There is new fundamental research that has led to new discoveries of the role of polyphenols in lignification and fermentation of maize cell walls (*41*) which may lead to more efficient ways to convert lignin to energy or to other useful products.

A continuing and future challenge is for technologies that reduce emissions, or capture and reuse emissions of greenhouse gases (CO_2, methane, NO_x). Gases such as CO_2 from combustion and fermentation, N_2O and NO_x from nitrogen fertilizers, CH_4 from ruminant metabolism and incomplete combustion, are examples of greenhouse gases ripe for reduction which could and should be associated with agricultural operations (*42*).

The phase-out of methyl bromide and other related fumigants and pesticides which are proven ozone depleters is an example of an area in which food producers need to be in the lead, rather than coerced into action by threats of increased regulation. As it turns out, the phase-out has accelerated the development of alternative nematicides and non-chemical methods for nematode control that are natural and environmentally friendly (*43*). Methane is a major greenhouse gas for which agriculture is an important source, with enteric fermentation emissions from cattle and other livestock making up nearly a third of emissions from the agricultural sector (*44*). Impressive reductions in methane emissions, without

10

sacrifice in milk or meat production, have occurred by using feeding rations low in roughage and independent of grazing (25, 45). The modern image of such practices that improve production efficiency with lowered greenhouse gas emissions is that they can entail aspects of 'factory farms' and potential for cruelty to animals. Developing such practices along with accompanying humane treatment of animals is a challenge for agriculture but one for which examples of success are emerging (46).

Concluding Thoughts

Many of these continuing advances in sustainability will be made with the increased use of bioenergy, biofuels, and biobased products , from farm, forest and food processing waste feedstocks (34, 47). Federal/state research and regulatory agencies are becoming increasingly active in setting standards for labeling items as sustainable (48, 49). Awards for green technologies are on the increase, and researchers and research funding sources have a role in promoting sustainability (50, 51). This is a time of much interest, ingenuity, and rapid pace of change in promoting sustainability, even as we struggle to define what it means.

References

1. Cobb, L. World population from 1800 to 2100, based on UN 2010 projections and US Census Bureau historical estimates; 2012. http://simple.wikipedia.Org/wiki/World_population (accessed June 11, 2013).
2. Food and Agricultural Organization (FAO). FAOSTAT3. http://faostat3.fao.org/faostat-gateway/go/to/browse/C/*/E.
3. Plunkett's Food Industry Almanac Market Research. http://www.plunkettresearch.com/food-beverage-grocery-market-research/industry-and-business-data (accessed February 13, 2013).
4. U.S. Department of Agriculture, Economic Research Service. *Census of Agriculture. Overview*; 2007.
5. JAFC Selects Virtual Issue: Food quality Traits for Sustaining Agriculture; 2011. http://pubs.acs.org/page/vi/2011/food-quality.html (accessed January 23, 2014).
6. Frankel, E. N. *J. Agric. Food Chem.* **2011**, *59*, 785–792.
7. Hafer, T. J.; Howard, L. R.; Prior, R. L. *J. Agric. Food Chem.* **2010**, *58*, 11749–11750.
8. Sharma, G. M.; Su, M.; Joshi, A. U.; Roux, K. H.; Sathe, S. K. *J. Agric. Food Chem.* **2010**, *58*, 5457–5464.
9. Argov-Araman, N.; Smilowicz, J. T.; Bricarello, D. A.; et al. *J. Agric. Food Chem.* **2010**, *58*, 11234–11242.
10. Finley, J. W.; Seiber, J. N. Maximizing the benefits of food; with the help of chemistry, we are eating safer, healthier, and more sustainable food than ever before. *Chem. Eng. News* **2011**, *89* (26), 46–51.
11. Everts, S. Chemicals leach from packaging. *Chem. Eng. News* **2009**, *87* (35), 11–15.

12. Cantrell, C. L.; Dayan, F. E.; Duke, S. O. *J. Nat. Prod.* **2012**, *75*, 1231–1242.
13. Casida, J. E. *Environ. Health Perspect.* **2012**, *120*, 487–493.
14. Lamberth, J.; Luksch, S.; Plant, T. *Science* **2013**, *341* (6147), 742–746.
15. Far Niente Sustainable Winery; 2013. http://www.farniente.com/WhoWeAre/SustainablePractices.html (accessed November 22, 2013).
16. SARE 2013. Sustainable Agriculture Research and Education, Western Region. Dixon Ridge Farms; September 3, 2013. http://www.westernsare.org/About-Us/What-is-Sustainable-Agriculture/Regional-Innovation/Dixon-Ridge-Farms (accessed November 22, 2013).
17. Dixon Ridge Farms. http://www.dixonridgefarms.com/home.html (accessed November 22, 2013).
18. Tareke, E.; Rydberg, P.; Karlsson, P.; Eriksson, S.; Törnqvist, M. *J. Agric. Food Chem.* **2002**, *50*, 4998–5006.
19. Zhang, G.; Huang, G.; Xiao, L.; Seiber, J. N.; Mitchell, A. E. *J. Agric. Food Chem.* **2011**, *59*, 8225–8232.
20. Mathews, J. M.; Watson, S. L.; Snyder, R. W.; Burgess, J. P.; Morgan, D. L. *J. Agric. Food Chem.* **2010**, *58*, 12761–12768.
21. Hengel, M.; Shibamoto, T. *J. Agric. Food Chem.* **2013**, *61*, 780–789.
22. Muttucumaru, N.; Elmore, J. S.; Curtis, T.; Mottram, D. S.; Parry, M. A. J.; Halford, N. G. S. *J. Agric. Food Chem.* **2008**, *56*, 6167–6172.
23. Friedman, M.; Levin, C. E. *J. Agric. Food Chem.* **2008**, *56*, 6113–6140.
24. Fröhlich, O.; Shibamoto, T. *J. Agric. Food Chem.* **1990**, *38*, 2057–2060.
25. Miller, G. D.; Auestad, N. *Int. J. Dairy Technol.* **2013**, *66*, 307–316.
26. Golan, E.; Krissoff, B.; Kuchler, F.; Calvin, L.; Nelson, K.; Price, G. *Traceability in the U.S. Food Supply: Economic Theory and Industry Studies*; Agricultural Economic Report No. 830; Economic Research Service, U.S. Department of Agriculture: 2004.
27. Seiber, J. N. *J. Integrative Agric.* **2012**, *11* (1), 9–13.
28. Williams, P. N.; Villada, A.; Deacon, C.; Raab, A.; Figuerola, J.; Green, A. J.; Feldmann, J.; Meharg, A. A. *Environ. Sci. Technol.* **2007**, *41*, 6854–6859.
29. U.S. Food and Drug Administration 2012. FDA Press Release, FDA releases preliminary data on arsenic levels in rice and rice products; September 19, 2012. http://www.fda.gov/newsevents/newsroom/pressannouncements/ucm319972.htm (accessed December 6, 2013).
30. Li, R. Y.; Stroud, J. L.; Ma, J. F.; McGrath, S. P.; Zhao, F. J. *Environ. Sci. Technol.* **2009**, *43*, 3778–3783.
31. Walmart takes new strides toward sustainability. http://www.thedailygreen.com/environmental-news/latest/walmart-sustainability-2013 (accessed December 12, 2013).
32. Walmart 2013 Global Responsibility Report. http://cdn.corporate.walmart.com/39/97/81c4b26546b3913979b260ea0a74/updated-2013-global-responsibility-report_130113953638624649.pdf (accessed December 12, 2013).
33. Worldometers; real time world statistics. World Population: Past, Present, and Future; 2013. http://www.worldometers.info/world-population/#pastfuture (accessed December 6, 2013).

34. Orts, W. J.; Holtman, K. M.; Seiber, J. N. *J. Agric. Food Chem.* **2008**, *56*, 3892–3899.

35. Duke, S. O. *J. Agric. Food Chem.* **2011**, *59*, 5793–5798.

36. Zhang, R.; Zhang, Z. U.S. Patent 6342378, 2002.

37. Nishiumi, S.; R. Mukai, R.; Ichiyanagi, T.; H. Ashida, H. *J. Agric. Food Chem.* **2012**, *60*, 9315–9320.

38. Zheng, Y.; Lee, C.; Yu, C.; Cheng, Y.-S.; Simmons, C. W.; Zhong, R.; Jenkins, B. M.; VanderGheynst, J. S. *J. Agric. Food Chem.* **2012**, *60*, 11128–11134.

39. Bilbao-Sainz, C.; Imam, S. H.; Franquivillanueva, D. M.; Wood, D. F.; Chiou, B.; Orts, W. J. *Basic Res. J. Soil Environ. Sci.* **2013**, *1* (3), 23–30.

40. Dhillon, G. S.; Kaur, S.; Brar, S. K.; Verma, M. *J. Agric. Food Chem.* **2012**, *60*, 7895–7904.

41. Grabber, J. H.; Ress, D.; Ralph, J. *J. Agric. Food Chem.* **2012**, *60*, 5152–5160.

42. Council for Agricultural Science and Technology (CAST). Intergovernmental Panel on Climate Change, CAST Task Force Report no. 142; October 2011.

43. Pesticide Action Network. Towards fumigant-free fields; 2013. http://www.panna.org/blog/towards-fumigant-free-fields (accessed December 18, 2013).

44. U.S. Environmental Protection Agency. Sources of Greenhouse Gas Emissions: Agriculture Sector Emissions; 2013. http://www.epa.gov/climatechange/ghgemissions/sources/agriculture.html (accessed December 16, 2013).

45. Hristov, N.; Oh, J.; Firkins, J. L.; Dijkstra, J.; Kebreab, E.; Waghorn, G.; Makkar, H. P. S.; Adesogan, A. T.; Yang, W.; Lee, C.; Gerber, P. J.; Henderson, B.; Tricarico, J. M. *J. Anim. Sci.* **2013**, *91*, 5045–5069.

46. Marcillac-Embertson, N. M.; Robinson, P. H.; Fadel, J. G.; Mitloehner, F. M. *J. Dairy Sci.* **2009**, *92*, 506–517.

47. Song, Y.; Wang, F.; Kengara, F. O.; Yang, X.; Gu, C.; Jiang, X. *J. Agric. Food Chem.* **2013**, *61*, 4210–4217.

48. U.S. Environmental Protection Agency, 2010. Information on Standards for Green Products and Services; 2010. http://www.epa.gov/epp/pubs/guidance/standards.htm (accessed December 12, 2013).

49. U.S. Department of Agriculture. BioPreferred website. http://www.biopreferred.gov/ (accessed December 12, 2013).

50. U.S. Environmental Protection Agency. P3: People, Prosperity and the Planet Student Design Competition for Sustainability; 2013. http://www.epa.gov/p3/ (accessed December 12, 2013).

51. U.S. Department of Agriculture. USDA Awards Research Grants to Ensure the Sustainability and Quality of America's Water Supply. News Release no. 0016.12; January 20, 2012. http://www.usda.gov/wps/portal/usda/usdahome?contentid=2012/01/0016.xml (accessed December 12, 2013).

Chapter 3

The New Inconvenient Truth: Global Contamination of Food by Chemical Pollutants, Particularly Heavy Metals and Metalloids

Tsananurayi Tongesayi[1,*] and Sunungurai Tongesayi[2]

[1]Department of Chemistry, Medical Technology and Physics, Monmouth University, West Long Branch, New Jersey 07764
[2]Walden University, 100 Washington Avenue South, Suite 900, Minneapolis, Minnesota 55401
*E-mail: ttongesa@monmouth.edu

There is a sense of vulnerability and inevitability in food contamination by environmental pollutants due to the current trend in environmental pollution, a situation that may be growing from the pressures of increased demand and limited resources in modern society. Modern society is presiding over an unprecedented overexploitation of the dwindling natural resources in its quest to maintain and advance human civilization and the demands of the increasing human population. Agriculture, industry, urbanization and technology have to match the demands of modern civilization and the accompanying population growth, but, unfortunately, the resources needed cannot match the demand. As a result, activities that may be deemed unethical are being practiced to the detriment of human health. Ironically, agriculture, industry, urbanization and technology, all man-made measures that are meant to enable society to cope with the demands of today, are primary polluters of the environment, including agricultural lands, and irrigation water. Food will inevitably be contaminated. Societal advancement has brought with it the idea of globalization, a concept that supposedly enhances the sharing of the dwindling resources, including food. Because pollution levels are not monolithic across the globe, globalization brings with it some unintended consequences, chief among them being the sharing of contaminated food. This article discusses the food contamination by heavy metal(loid)s, the health effects of heavy metal(loid)s, and possible solutions to a problem that is escalating to one of the major challenges of humanity in the 21st century. Proactive and collective pragmatic approaches are urgently required to protect human health.

Introduction

Humans, like most forms of life, live not to eat but rather eat to live. This makes food a universal life-saving need and resources that are required to produce the food especially important. Food production needs land and water, resources that form part of the Earth's natural endowment. The two resources, land and water, are laden with natural chemicals. In fact, they are chemicals by all means. All chemicals are potentially harmful. Some are downright toxic while some are essential for life in trace levels, beyond which they are harmful. The latter class constitutes essential poisons. Plants that produce human food take up chemicals from the soil and water. Some of the chemicals are essential for plant growth and food productivity while some not. The latter can be harmful to the plant and/or the human consumer. Besides the natural spatial and temporal distributions of chemicals in the environment, which include the land and water, man-made activities, commonly referred to as anthropogenic activities, contribute significantly to the redistribution of chemicals, both natural and synthetic, in the natural environment. Anthropogenic activities that contaminate the land and water are some of the key drivers of societal advancement and modern civilization. These include industrialization, urbanization, and extensive use of agrochemicals to increase food productivity. The ultimate consequence of the foregoing is the inevitability of chemicals inadvertently getting into human food. To protect human health, quality must be one of the key defining parameters in food production.

The earth's land endowment is a limited resource while human population is a natural variable. The demand for food obviously increases with an increase in population size. In modern society, humans are living longer than before, and more humans are being born with greater chances of surviving into adulthood than ever before, all thanks to modern civilization. As a matter of fact, the world population is projected to reach 9.1 billion by 2050 from the current 6.5 billion, while essential resources such as land and water are dwindling (*1–3*). To meet the food requirements while maintaining societal advancement and civilization, the growth in world population has to be matched by growths in agriculture, industrialization and urbanization. Ironically, these anthropogenic activities are the primary polluters of the natural environment that include agricultural lands and irrigation water, and hence human food. It is, therefore, logical to project an inevitable increase in the contamination of food as the world population increases if measures are not put in place to control environmental pollution. At stake is human health, and something drastic has to be done, obviously, by humans.

Modern civilization has brought with it the concept of global citizenry. Humans now live in what is termed the global village; nearly every geographic part of the world has the qualities of a "melting pot", a term used to describe a population or community with people of all backgrounds, races and ethnicities. The melting pots are also characterized by a variety of tastes and preferences, chief among them being food. As a result, there is always a need for a variety of foods among other consumer goods in almost every corner of the world (4). Thanks to societal advancement, food, just like people, can easily be moved from one part of the world to the other in any quantity at unprecedented speeds and frequencies, and, at times without having to preserve the perishables, they arrive just as fresh. Food does not have to be produced locally. The new norm is "produce locally and serve globally". It is no longer just quality and quantity that must characterize food consumption; variety has been added to the mix.

The globalized food market and the need for diverse foods may come at a cost; the unintended consequences, especially health risks associated with contaminated food. Within the global village, natural environmental conditions, agricultural practices and policies that govern environmental pollution from anthropogenic activities are not monolithic. Some regions of the globe are naturally more polluted than others and in other regions agricultural lands, and irrigation waters are more prone to chemical contamination as a result of the laxity or non-existence of laws governing anthropogenic activities such as agriculture, mining and industry with regard to environmental pollution. Research has shown that some of these land- and water-polluting anthropogenic activities are more prevalent in regions that are some of the major producers of food for the world population (5, 6). As a result, regions of the world where the agricultural soils and irrigation waters are contaminated could be acting as conduits for toxic chemical exposures to the rest of the world population. This essentially means that consumers in both the polluted and non-polluted regions of the world are equally at risk from the health effects of environmental chemical pollutants through food.

According to the National Institute of Environmental Health Sciences (NIEHS), exposure to environmental contaminants plays a role in 85% of all diseases (7), particularly chronic diseases such as obesity, diabetes, heart disease, and cancer. The postnatal environment has long been known to play a crucial role in determining human susceptibility to disease, but there is now a growing body of scientific evidence that suggests a link between prenatal exposure to chemical toxicants and the development of chronic diseases such as obesity, diabetes, heart disease, and cancer, later in life. This, according to scientific data, is a result of epigenetics, the alteration of gene expression without changing the gene sequence. It is now widely accepted that exposure to heavy metals and metalloids (metal(loid)s), in particular, is one of the primary causes of human cancers via both genetic and epigenetic mechanisms (8). With regard to diabetes, existing literature specifically links toxic heavy metal exposures to type-2 diabetes (9). The heavy metal(loid)-induced epigenetic mechanisms have been shown to involve heritable changes in gene expression without alteration of the DNA sequence, achieved via changes in the structure of chromatin. Genes are expressed when chromatin is in an extended state and are inactivated when chromatin is condensed. Both the two chromatin formations are controlled by reversible epigenetic patterns of histone modification and DNA methylation. Overwhelming scientific evidence support the fact that heavy metals, such as nickel (Ni), lead (Pb), arsenic (As), cadmium (Cd), copper (Cu), and chromium (Cr), can cause direct modification of the epigenetic state of the genome via DNA methylation, histone modification, and the expression of small, non-coding RNAs (*10*). According to *the fetal basis of adult disease or windows of susceptibility* concept, environmental factors, food and behavioral changes that may have minimal adverse effects in adults may adversely impact the development of a fetus and may induce chronic health effects on a child even in adulthood (*7*). This shows that these environmental pollutants do not have to be present in large amounts to cause adverse health effects, making it a significant health hazard for expecting mothers to be exposed to levels of environmental pollutants that are considered safe by current standards.

The principal source of chemical exposure to humans has generally been accepted to be drinking water. As a result, most of these chemical pollutants, particularly toxic heavy metals and metalloids, have established maximum contaminant levels (MCL) in water that are enforced locally by individual nations and globally by organizations such as the World Health Organization (WHO). Food has never been considered a significant threat to human health as a source of chemical toxicants until recently. This may explain why there are no established safe limits or MCLs for toxic chemicals in food (*11, 12*). However, the dynamics are changing, and food is emerging as one the significant sources of chemical exposure to humans (*13, 14*). Because of the importance of food to life, the devastating health effects of chemical toxicants and the inevitability of the potential contamination of food by environmental chemical pollutants, proactive and pragmatic solutions need to be instituted. Reactionary as well as the head-in-sand approach will put human health at risk. Food contamination puts human health in a direct and immediate danger, and if not addressed proactively and collectively, will soon escalate into one of, if not, the major global challenges of the 21st century.

Most, if not all, countries are importing food from other countries to meet diversity needs (*4*) and maybe local shortages. The United States, for example, imports a variety of foods from around the world to meet the demands of its ethnically diverse and economically well-to-do population for diversity, quality and convenience in the food they eat (*4*). According to the US Department of Agriculture (USDA) the food imports are also driven by seasonal and climatic factors, especially foods such as fruits, vegetables, and tropical foods such as cocoa and coffee. Intra-industry trade also accounts for a portion of the food imports into the country (*4*). The following is a comprehensive list of foods that the US imports for its diverse population: live meat animals; meat and meat products; fish and shellfish; dairy products; vegetables and vegetable preparations; nuts and nut preparations; coffee, tea, and spices; grains, grain products, and bakery foods; vegetable oils and oilseeds; sugar, sweeteners, and confections; cocoa products and chocolate; sauces, essence oils, and other edibles; wine, beer, and other beverages (*4*).

Cereal grains and rice in particular are considered the major food sources of exposure of toxic heavy metal(loid)s to humans (*12, 13*). The rice plant efficiently accumulates heavy metal(loids) compared to other cereals, and most terrestrial-based foods (*15*, *16*), significantly enriching the metals in the grain and other parts of the plant at levels several-times higher than the soil (*17*). The enrichment of metal(loids) in the rice plant and the fact that rice is a staple food for over half of the world population make it an ideal candidate for the evaluation of the potential threat posed by food as a source of toxic heavy metal(loids) to humans. In 2012, the US imported 10,184,400 metric tons (mt) of total grains and products with a total value of $9,083.1 million (m). This represented 52% increase compared to the 1999 imports of total grains and products. The major suppliers of US total grains and products in 2012 in terms of the dollar amount spent were: Canada (50%), Mexico (11%), Thailand (6%), Italy (4%), Brazil (3%), India (3%), Germany (3%), and China (2%). The rest of the world supplied the remaining 18%. With regard to rice and flour, the US imported 625.8 mt with a total value of $659.5m in 2012, an increase of 73% compared to the imports in 1999. The major suppliers based on the dollar amount spent were Thailand (65%), India (21%), Vietnam (4%), Pakistan (3%), and Italy (2%), with the rest of the world supplying 5%. Apart from grains, fruits and fruit juices have also been identified as some of the primary sources of toxic heavy metal(loid)s, particularly as to humans (*14*). Last year, the US imported 8193.5mt of total fruit and preparations worth a total of $4,793.5m with the major supplier being Mexico (20%). In the prepared fruit and fruit juices categories, China was the major supplier, accounting for 27% and 35% of the imports respectively.

The EPA's Toxic and Priority Pollutant metal(loid)s list include antimony (Sb), arsenic (As), beryllium (Be), cadmium (Cd), chromium (Cr), copper (Cu), lead (Pb), mercury (Hg), nickel (Ni), selenium (Se), silver (Ag), thallium (Tl) and zinc (Zn). Some of these priority pollutants are non-essential and toxic while others are essential but toxic above certain thresholds. A substantial amount of research on the contamination of food by toxic metal(loid)s has focused on As, Cd and Pb. As a result, this article focuses on three of these metal(loid)s, reviewing their effects on human health, their presence in food and potential sources. Manganese, even though it is not listed as one of the priority pollutants, is one of the most devastating essential poisons, and together with another essential poison, Zn, will also be discussed in this article. The article will also suggest and critically evaluate possible solutions to food contamination.

Heavy Metal(loid)s and Human Health (As, Cd and Pb)

Arsenic

Arsenic is a metalloid that occurs naturally in the environment. It can also get into the air, water and soil as a result of anthropogenic activities such as agriculture and industry. Arsenic is a classified neurotoxin and carcinogen (*18*). The inorganic forms of As, arsenite (As(III)), and arsenate (As(V)), i.e. mineral As without an organic moiety, are the most toxic of all forms of the metalloid, with As(III) being more toxic than As(V). The metalloid has been shown to cause cellular proliferation, apoptosis, and differentiation, and metabolic changes, as well as acting as a co-carcinogen. After realizing the devastating human health effects of As, the US Environmental Protection Agency (U.S.EPA) lowered its maximum contaminant level in drinking water from 50 ppb to 10 ppb in January 2006 (*19*).

Humans are exposed to As through the consumption of contaminated drinking water, soil, and food, or through inhalation of contaminated dust and air (*10, 18*). Exposure to As is widely accepted to be associated with lung, bladder, kidney, liver, and non-melanoma skin cancers, as well as cardiovascular diseases and diabetes (*10*). Arsenic exposure causes a decrease in the expression of the phosphatase and tensin homolog (PTEN) tumor suppressor gene, leading to an increase in cancer stem cells and cancer development. The metalloid induces tumorigenesis through several mechanisms that include oxidative stress, genotoxic damage, and chromosomal abnormalities, and, recently, epigenetic mechanisms (*18*). The epigenetic mechanisms involve the alteration of methylation levels of global DNA and gene promoters; histone acetylation, methylation, and phosphorylation; and miRNA expression (*18*). Studies have shown extensive methylation of specific genes in people exposed to As through drinking water (*10*). Arsenic is the only known environmental pollutant that induces changes in all three epigenetic markers, that is, DNA methylation, histone modifications and expression of noncoding RNAs (*20*). The epigenetic manifestations of arsenic also include the acceleration and exacerbation of the formation of plaque in arteries, increasing the risk of cardiovascular diseases (*10*). Sufficiently high levels of the metalloid have been associated dyspigmentation, keratosis, peripheral vascular diseases, reproductive toxicity, and neurological effects (*18, 20*).

Prenatal and early childhood exposure to As at sufficiently high levels has been linked to the development of chronic health effects such as cancers of the bladder, lung, and skin, as well as cardiovascular and respiratory diseases in early life or later in life (*21–23*). Research by Liaw et al. (*24*) reported a higher incidence of liver, lung, and kidney cancer in adulthood in Chilean children who were exposed to high levels of naturally occurring As via drinking water. Another recent study in Japan by Yorifuji (*25*), reported higher mortality rates for skin and liver cancer in infants that were fed arsenic-contaminated milk powder. In addition to the development of chronic diseases later in life, research has also showed that pre-natal exposure to low or moderate levels of As has been shown to lead to neurological and cognitive dysfunction, as well as memory and learning impairments (*20, 26*). Epidemiology and models involving animals also indicate a link between utero exposure to As and fetal health and low birth weight (*22, 27*). Exposure to As has also been shown to cause nausea, vomiting, diarrhea, abnormal heart rhythm, blood vessel damage, and a pins and needles sensation in hands and feet (*28*).

Cadmium

Cadmium is a heavy metal that occurs in the natural environment in lead, copper, zinc, and other metal ores, and is widely used in the chemical industry. Anthropogenic activities such as cadmium-emitting industries, fossil fuel combustion, and waste incinerators are mainly responsible for the contamination of agricultural soils, water, and air. Cadmium is classified human carcinogen and has been linked to cancers of the pulmonary system, liver, bladder, stomach, renal, and the hematopoietic system (*10*). Long term exposure to low levels of Cd has been linked to kidney and cardiovascular disease, fractures, and cancer (*29*). Cadmium exposure is also associated with an increase in cancer susceptibility in type-2 diabetes, and a prolonged exposure to the metal is associated with pre-diabetes, diabetes, and overall cancer mortality which is sex-dependent in some types of cancer (*30*). A study by Ciesielski et al. (*31*) suggests a link between higher cumulative Cd exposure to levels that have been considered safe and subtly decreased performance in attention- and perception-related tasks by adults especially those whose primary exposure is food. These levels of Cd are reportedly common among adults in the US. Another published scientific study suggests that children and teens exposed to Cd levels that were considered safe and common among US children are more likely to develop learning disabilities (*32*). A recent study conducted by Zhang et al. (*33*) in Xuanwei and Fuyuan, two of the high cancer incidence areas in the Yunnan province of China, reported significantly higher levels of Cd and Ti in the diet of people in the cancer-stricken areas than in control groups.

Although the mechanism of Cd toxicity is largely unknown, current evidence suggest epigenetic mechanisms that involve both hypo- and hyper-methylation of DNA (*10*). The DNA methylation status influences gene transcription and Cd exposure can induce cancer if the genes affected are the tumor suppressor genes (*10*).

Food and tobacco smoking are considered the main sources of exposure to humans (*29*). Tobacco smoking is considered the source with the greatest potential for above-average exposure of Cd to smokers (*32*). With regard to the human diet, cereal grains, particularly rice, are considered to be the principal sources of Cd (*34, 35*). A recent study conducted in Shanghai, China reported that vegetables and rice were the main sources of dietary cadmium intake followed by tobacco which accounted for about 25% of the total dietary cadmium exposure (*36*). Other sources of Cd exposure to humans include contaminated drinking water, and inhalation of contaminated dust and air.

Lead

Lead occurs in trace levels naturally in the environment. However, anthropogenic activities have changed the metal's natural spatial distribution in the environment to the extent that it is now being found in unsafe levels in environmental compartments where it is readily accessible to humans. For instance, Pb is now being found in unhealthy levels in drinking water, soil, and agricultural products. The principal anthropogenic sources of Pb in the environment include mining activities, lead-based paints and pigments, ammunition for hunting, solder, Pb weights, and bearing metals, production of iron and steel, lead acid batteries, stabilizers in PVC and cosmetic formulations. The extensive use of lead-based pesticides such as lead arsenate and leaded gasoline, even though they are now banned in the US and other countries, contributed, in a significant way, to the loading of Pb in the environment.

Lead is a non-essential neurotoxic heavy metal that primarily affects the central nervous system, the renal system, skeletal system, and hematopoietic systems. The pre-, peri-, and postnatal exposure to Pb causes severe and mostly irreversible mental retardation, learning disabilities, and sometimes even coma and death in children (*10*). Pb manifests its toxicity by causing oxidative stress and the disruption of cell signaling and neurotransmission pathways as well as influencing the methylation of DNA methylation (*10*). Scientific literature also suggests that exposure to Pb in early life can induce epigenetic mechanisms which increases susceptibility to disease later in life (*10*). According to Cory-Slechta et al. (*37*) exposure to Pb during the early stages of development may alter the hypothalamic-pituitary-adrenal system which regulates the functions of many body organs, a phenomenon that may explain the Pb-induced increased risk of hypertension, cardiovascular disease, diabetes, schizophrenia, and neurological diseases in adult life.

The subtle irony of Pb poisoning is that affected children may not show distinct physical signs. Also, the gastrointestinal absorption of lead declines with age, meaning that children absorb more lead than adults. Because of the detrimental health effects of lead poisoning and its physical elusiveness in children, the Centers for Disease Control and Prevention (CDC) now recommends screening children in high risk areas or populations for Pb exposures (*38*). In adults, Pb poisoning increases blood pressure and cardiovascular diseases as well as inducing calcium deficiency by replacing the calcium in bones. The heavy metal is also associated with reproductive defects. The major targets for Pb after its gastrointestinal absorption include blood plasma, nervous system, and soft tissues. It is, however, subsequently redistributed and accumulates in bone. About 75% to 90% of the total Pb in the human body is stored in bones and teeth. A report by the International Agency for Research on Cancer (IARC) in 1987 concluded that there was no sufficient evidence that links Pb exposure to human cancers, but recent research suggest otherwise, especially with regard to lung and stomach cancers (*39*).

The main exposure sources of Pb to humans used to be contaminated soil, household dust, drinking water, lead crystal, lead-glazed pottery, and some inexpensive metal jewelry (*40*), but food is emerging as one of the major sources of the heavy metal. The heavy metal is also present in food cans and metal plumbing. The devastating health effects of Pb led to the banning of lead paint, lead pesticides and leaded gasoline. The bans reduced direct exposure of Pb to humans, albeit temporarily because the overall consumption of Pb is reported to have actually increased primarily due to the increased production of lead-acid batteries (*39*). The bans were, however, cited by the National Institute of Environmental Health Sciences (NIEHS) as the main reasons why lead poisoning in American children had decreased by about 86% since the late 1970s (*40*). However, the peeling of lead paint from old homes and other emerging sources of Pb still pose significant threats to human health. Recent studies by the CDC showed that 890,000 U.S. children aged 5 or younger have elevated blood Pb levels, with more than 20 % of African-American children living in housing built before 1946 having elevated blood Pb levels. The source of the Pb was thought to be Pb paint in the old houses and contaminated dust and soil (*38*).

Essential Elements and Human Health (Mn and Zn)

The amounts of essential elements in foods have been known to meet health requirements without any health risk of inadvertent overdoses. This may be the reason why regulatory bodies and scholars tend to focus on non-essential toxic elements when testing for inorganic environmental chemical contaminants in food. However, present trends in environmental pollution resulting from extensive and intensive industrialization, urbanization, mining and agriculture, suggest potential risks from essential metal(loid) toxicity via the food chain. Both essential and non-essential elements are increasingly getting into the food chain as a result of the extensive use agrochemicals, and the use of contaminated water, raw sewage, and untreated industrial effluent to irrigate crops (*41, 42*). Comprehensive and holistic testing approaches for chemical contaminants in food should be the norm, especially considering that the essential elements are as a matter of fact essential poisons (*43*), if human health has to be protected. They are essential but toxic above certain limits. The eating of contaminated foods that are considered non-dietary sources of the essential but toxic elements may result in inadvertent overdoses either from the food alone or from the food and supplements that consumers may be simultaneously taking.

Manganese

Manganese occurs naturally in the soil, but, just like other heavy metal(loid)s, anthropogenic activities such as agriculture, industry, mining and waste management are increasing its levels in water, air and soil to the extent of becoming potential health hazards, particularly through the food chain.

Manganese is an essential but neurotoxic element with long-lasting and potentially irreversible effects. Levels beyond the recommended daily intakes can result in a Parkinson's disease (PD)-like syndrome and children's exposure can severely affect neurological development (*43*). Its toxic effects occur in the respiratory tract and the brain, and symptoms of toxicity may include hallucinations, forgetfulness and nerve damage. An overdose of Mn is known to be associated with lung embolisms and bronchitis. In men, Mn can cause impotence from prolonged exposure. Mn poisoning also causes a syndrome whose symptoms may include schizophrenia, dullness, weak muscles, headaches and insomnia. On the other hand, a deficiency of Mn is known to cause negative health impacts that include obesity, glucose intolerance, blood clotting, skin problems, lowered cholesterol levels, skeleton disorders, birth defects, changes of hair color as well as neurological symptoms (*44*).

Manganese uptake by humans is mainly through foods, such as spinach, tea and herbs. Other foodstuffs that are reported to contain Mn are grains and brown rice, soy beans, eggs, nuts, olive oil, green beans and oysters (*44*). It has to be noted that the proportion of Mn in brown rice is considered insignificant enough that the USDA does not even list it on the nutrient composition of brown rice.

Inadvertent exposure of Mn to humans occurs via the inhalation of contaminated dust and fumes.

Zinc

Zinc is another example of an essential poison that is obtained by humans primarily from protein-rich foods such as beef, lamb, pork, crabmeat, turkey, chicken, lobster, clams and salmon. An overdose of Zn causes stomach cramps, skin irritations, vomiting, nausea and anemia (45). At sufficiently high levels, the essential poison is reported to cause damage to the pancreas, disrupt protein metabolism, and to cause arteriosclerosis. Zinc can be passed from the mother to the fetus through blood and from mother to child via breast milk, and hence can be a severe threat to both the unborn and newborns (45). The deficiency of Zn in the human body is associated with the loss of appetite, taste and smell. Insufficient Zn also causes skin sores, slow wound healing and birth defects.

Heavy Metal(loid)s in Food: Potential Sources

As pointed out earlier, food and beverages such as rice, grains, and juices are emerging as some of the primary sources of human exposure to heavy metal(loid)s (13). Rice is particularly of significant concern because rice products are used as ingredients in prepared foods some of which may not show an obvious rice presence. For instance, organic brown rice syrup is used as a sweetener in place of high-fructose corn syrup in organic food products. In fact, work done by Jackson et al. (11) reported high concentration of As in the organic brown rice syrup samples that they analyzed. They also reported As levels that were six times higher than the US EPA drinking safe limits in organic toddler milk formula samples containing organic brown rice syrup as the main ingredient. High As levels were also reported in cereal bars and high-energy foods that were prepared with organic brown rice syrup. Similar products without organic brown rice syrup had lower levels of As. Carbonell-Barrachina et al. (46), in a recent study reported high levels of As in pure baby rice sample from UK and U.S. compared to those from China and Spain.

In September 2013, the FDA released a comprehensive list of As test results for more than 1,300 rice and rice food products (12). The concentration of inorganic As (iAs) in rice samples ranged from 3.5 to 7.2 µg/serving; infant formula 0.1 µg/serving (148 g serving); infant cereal 1.8 µg/serving; toddler cereal 1.5 µg/serving. A study on rice samples collected from Thailand and some Asian countries reported As levels that ranged from 22.51 µg kg^{-1} to 375.39 µg kg^{-1} (47).

A recent study that measured the levels of Cd in rice samples from 12 countries on four continents reported the highest levels of the metal in samples from Bangladesh and Sri Lanka. The calculated weekly intakes of Cd from rice where considered unsafe (*34*). Carbonell-Barrachina et al. (*46*), in a recent study, also reported high levels of Cd in pure baby rice samples from China compared to those from Spain, the UK and the US. A study by Gupta et al. (*48*) in wild rice reported elevated levels of As and Pb which they thought was a result of atmospheric deposition.

Tongesayi et al. (*49*) measured levels of manganese and zinc in rice and calculated the daily bioaccessible levels of the two elements. The daily bioaccessible levels were significantly higher than the recommended daily intakes in most of the samples. Rice is not considered a dietary source of Mn and Zn, and on nutritional labels on rice packaging, the two essential elements are not listed, showing that they constitute an insignificant portion of the nutrients in the grain. Therefore, consumption of contaminated rice may potentially result in an overdose, considering that consumers may be taking supplements as well as foods that are recommended as sources of the essential elements. Because exposure from various sources is additive, lower levels than recommended daily intakes in one source may not guarantee safety from a particular chemical toxicant in cases where there are multiple exposure sources.

It has long been accepted that the extensive use of agrochemicals was the principal anthropogenic source of heavy metals and metalloids in the food chain. However, the dynamics of food contamination are changing, primarily because of unethical waste-management practices that may a result of the generation of both hazardous non-hazardous wastes in both physically and economically overwhelming quantities on a daily basis; unethical agricultural practices that may be a direct consequence of the increased demand for food in the face of dwindling resources such as water and land; and the unprecedented levels of industrialization and urbanization that are a direct result of the need to maintain an advanced civilization in an era where population is increasing and resources are dwindling. It is becoming increasingly inevitable for environmental chemical pollutants to enter the food chain.

Hazardous waste, particularly electronic waste (e-waste) has become a huge problem for nations, and its movement across borders is presenting significant environmental challenges to receiving nations. The shipment of e-waste is severely contaminating local environments that include farmlands and irrigation water in the receiving countries. Food will almost obviously be contaminated, and in a typical what-goes-around-comes-around fashion, the poor countries will return the "favor" from the rich countries in kind; in the form of contaminated food, because of the globalized food market. The consequence is that all humans, rich or poor, are put at risk. Unfortunately, food becomes the ultimate equalizer with regard to human health, indeed a very disturbing situation.

In most developing countries, the shortage of water, land, increasing industrialization and urbanization have resulted in crops being irrigated with untreated industrial and sewage effluents and freshwater contaminated by leachates from landfills and acid mine drainage. In some cases, landfills, mines, industries, highways and urban centers are inseparable from agricultural lands. The ultimate consequence is human health risk from food contamination by environmental pollutants. Such agricultural practices appear to be widespread in some countries that are some of the major producers of food for the world population (*42*). To make matters worse, the practices appear to be on the rise despite numerous research data pointing to their potential negative impact on human health. Below are typical examples of the impacts of e-waste and food-contaminating agricultural practices on the environment and human health.

Electronic Waste

Waste management is a daunting and costly task that presents unprecedented challenges to any local, national, or international organization. Waste management needs lots of land, and land is a limited resource. Population growth, accompanied by the large volumes of high-turnover technological items and other requirements for modern civilization, has resulted in the generation of unmanageable volumes of waste, both hazardous and non-hazardous. Because of the resulting shortage of land and the cost of recycling, large volumes of waste, particularly hazardous waste, are crossing oceans and borders, with final destinations being primarily developing countries (*50*). This is done on the pretext of recycling, yet it is a form of waste management by rich nations (*51*).

A typical example is the movement of e-waste. E-waste includes discarded computers, computer monitors, television sets, and cell phones, and is fast becoming one of the largest growing streams of waste globally. Its management and recycling is extremely challenging due to the complex nature of its composition and a lack of well-established recycling methods (*52*). Most of these electronic items contain toxic and carcinogenic heavy metals such as Pb, Hg, Cd, hexavalent Cr, Ni, Sb and organic toxicants and carcinogens such as poly-brominated di-phenyl ethers and polychlorinated biphenyls (*53*). It also contains valuable metals such as Cu and the platinum group metals (*53*). Recycling of e-waste to recover reusable valuable materials with minimal environmental impact is very expensive (*53*). Because of the prohibitive cost of recycling and managing e-waste, rich countries export unknown but large quantities of e-waste to poor countries in spite of the Basel Convention (*53*).

For example, it is reported that e-waste from the United States and other developed countries is exported mainly to Asian and African countries (*54*, *55*). There is, however, no evidence that such activities are being sanctioned by governments in both the source and destination nations. One of the major consequences is contamination of local environments including water, soils and crops in the receiving countries (*53*). Ironically, the food that is produced in the e-waste-contaminated countries may not be consumed by local people only, but some of it may be exported to other countries around the globe that include the same developed countries that shipped out the hazardous waste. The developed countries may be benefiting in the short-term by avoiding costly waste management but may suffer long-term effects of food contamination.

People involved in e-waste recycling have been reported to suffer adverse health effects through skin contact and inhalation while the rest of the people within the communities are exposed to the toxicants through smoke, dust, drinking water, and food (*53*). In a review article by Chen et al. (*57*) it was noted that environmental exposure to Pb, Cd, Cr, poly-brominated di-phenyl ethers, polychlorinated biphenyls, and polycyclic aromatic hydrocarbons was prevalent at high concentrations in pregnant women and young children, particularly in developing countries, because of informal and primitive e-waste recycling practices. The review also noted that developmental neurotoxicity was a serious concern in developing countries. Recycling techniques in most of the destinations for e-waste include burning and acid-dissolution, in most cases without any measures to protect the environment and human health (*53*). The burning of e-waste produces an array of organic toxicants such as dioxins, furans, polycyclic aromatic hydrocarbons, polyhalogenated aromatic hydrocarbons, and HCl (*53*).

Despite the large amount of scientific data that shows the negative environmental impacts of e-waste in developing countries where recycling practices are mainly informal and primitive, e-waste shipments to developing countries are increasing at unprecedented rates (*56*). According to a review by Robinson (*53*) in 2009, the global production of e-waste stood at 20-25 million tons per year, with the bulk of it having been produced in Europe, the United States, and Australasia. Some researchers in India, for instance, have reported an acute accumulation of e-waste from developed countries, which they argued was threatening the environment and human health and blamed it on the ambiguity in national and international laws that govern movement of waste across borders (*51*). According to Gupta et al. (*58*), about 80 percent of the e-waste that is generated in the US is exported to India, China and Pakistan under the name of charity with the situation in India being worsened by the large amounts of e-waste that the country is also now generating, while only recycling 3% of the total. The remaining 97% is informally recycled by people who work with bare hands under unhygienic conditions and without protective gear.

In a study by Tang et al. (*59*) in the Taizhou area, China, it was shown that agricultural soils were being contaminated by inorganic (Cr, Cd, Pb, Zn, Cu, and Ni) and organic (PAHs, and PCBs) pollutants as a result of e-waste recycling. Another study by Fujimori et al. (*60*) in Metro Manila in the Philippines reported an enrichment of Ag, As, Cd, Co, Cu, Fe, In, Mn, Ni, Pb, and Zn in soil and dust surface matrixes from formal and informal e-waste recycling sites. The levels of the metal(loid)s at informal sites were higher than those at formal sites and were similar to levels that were found at informal recycling sites in other Asian countries. Another study in Bangalore and Chennai in India by Ha et al. (*61*) that was conducted at e-waste recycling sites reported levels of Cu, Zn, Ag, Cd, In, Sn, Sb, Hg, Pb, and Bi in soil that were higher than at reference sites. The levels of Cu, Sb, Hg, and Pb in soils at some of the e-waste sites exceeded screening values proposed by U.S. EPA.

In China, a study by Luo et al. (*62*) in agricultural soils and vegetables within and around e-waste sites reported high levels of Cd, Cu, Pb, and Zn with mean values of 17.1, 11,140, 4500, and 3690 mg/kg, respectively. In vegetables, the levels of Cd and Pb that were reported exceeded the maximum level permitted for food in China. A study conducted by Fu et al. (*63*) in rice in an e-waste area in China reported levels of Pb and Cd that exceeded the national maximum allowable concentration and the FAO/WHO tolerable daily intakes. The levels of Pb and Cd found were also higher than levels found in commercial rice samples showing that the e-waste site was the most probable source of contamination.

Agricultural Practices and Food Contamination

A review by Murtaza et al. (*42*), shows that the use of sewage effluent, treated or untreated, is rampant, particularly in the developing world and is ongoing despite the amount of scientific literature and electronic media reports that show the environmental and public health ills of the practice. Most, if not all, research is showing significant enrichment of both toxic and essential metals as well as organics and pharmaceuticals in agricultural soils and crops. According to the literature, this practice is being necessitated by water shortages and the perennial droughts in some regions of the world, as well as the nutritional value of sewage (*41, 42*). Plant nutrient levels will obviously increase but at the expense of soil contamination and, ultimately, public health.

The composition of sewage has significantly changed over the years thanks to societal advancement and modern civilization. It has moved from being mainly carbohydrate-based domestic waste to a complex mixture composed of heavy metals, essential elements, persistent organics and pharmaceuticals, in addition to polymeric carbohydrates and nutrients such as phosphates and nitrates (*64–66*). Some developed countries realized the public health hazards of the practice and stopped it, but not after having used sewage water to irrigate their farmlands for over a century (*64*). Chemical measurements in the soils previously irrigated with sewage before the practice was stopped still show elevated levels of toxic chemicals, particularly heavy metals and metalloids that are a potential threat to human health (*67*). In some of the developed countries where sewage effluent is no longer or not being used to irrigate crops, sewage sludge, commonly referred to as bio-solids, which can be highly contaminated by toxic chemical pollutants, is being used as a fertilizer, soil amendment and or for land reclamation (*68*). The sewage sludge is used in these ways on the pretext of recycling, yet the truth of the matter may be the cost of managing large volumes of sludge and the shortage of land for landfills (*68*).

Most of the work done and published in the scientific literature has raised serious concerns about the use of sewage and industrial effluents to irrigate crops, and recommended the banning of the practice. However, the amount of literature on the research concerning the use of sewage as irrigation water varies from country to country, with very little to none in some countries where the problem exists. The efforts of scientists in some of the countries where a substantial amount of research has been conducted are being augmented by local media outlets that have taken it upon themselves to expose the health hazards of the practice, and to educate their communities, as well as calling upon the responsible authorities to act.

Credit should be given to scientists in India and China, and the media in China for their devotion to the issue of food contamination by toxic chemicals from anthropogenic activities. Coincidentally, these two countries are some of the countries that play a pivotal role in the global village with regard to food provision.

According to credible scientific literature, India has some of its agricultural crops being irrigated with untreated industrial wastewater and sewage water (*69–76*). Also, some of its industries, urban centers, solid waste sites and highways are located within the vicinity of agricultural lands (*75, 77–79*). According to published research conducted in the country, these practices, together with the extensive use of agrochemicals, are contributing to the unexpectedly high levels of toxic heavy metal(loids) in India's agricultural foods. Several researchers have reported elevated levels of heavy metal(loids) in the industrial effluents and sewage waters as well as the agricultural soils in the country. Analyses of heavy metal(loids) in rice and other food crops in the country have reported levels that were higher than recommended limits, and that correlated positively with levels in the soils (69, 71–73, 78, 79). Research on crops that were irrigated with uncontaminated water on agricultural lands away from landfills, highways, and industries had very low to undetectable levels of the heavy metal(loid)s (75, 78–81). This shows that contaminated water, untreated domestic and industrial effluents, landfills, highways and industry are the most probable sources of food contamination in the country. It is not clear whether such practices are uniform across the country or are confined to particular regions and whether the food produced under these conditions is all consumed locally or some is exported to other countries. What is clear is that the practices are potentially putting human health at risk and need to be remedied.

In China, the media has joined efforts by scientists to publicize the public health problem that is associated with food contamination from anthropogenic activities. As in most developing countries, crops in some parts of the countries are reportedly being irrigated using raw sewage water (82–84), and irrigation water is reportedly contaminated with heavy metals from mining and industrial discharges (85). Agrochemicals, solid waste, and highways have also been cited by researchers as some of the sources of contamination of agricultural soils and crops in this region (86, 87). Many studies have shown that some agricultural soils and crops in this region are heavily contaminated with heavy metal(loids), with the levels in crops being positively correlated with those in the soils (88–91). Most of these studies reported significant enrichment of heavy metal(loids) in crops relative to the levels in the soil.

Regular headlines in Chinese media about heavy metals in rice are an indication of concern about the extent of food contamination in the country and the magnitude of the threat to human health. The following are some of the recent headlines that were run in some Chinese electronic media about heavy metals in rice: *Harmful heavy metals found in Hunan-produced rice* (92); *Heavy Metals Tainting China's Rice Bowls* (93); *China rice laced with heavy metals*: (94); *The Crisis of Tainted Rice: Soil pollution is impacting one of China's long-cherished staples. What's to be done?* (95); *New Rice Contamination Reported in China* (96); *Heavy Metal Pollution Threatens China's Rice Industry* (97) *Heavy metal fears devastate market for rice farmers* (98); *China Grown Rice Defiled With Heavy Metals and Chemicals* (99). These headlines by science journalists in China show that the contamination of food is a well-known and worrisome trend in China. Sewage water and industrial effluents that are used to irrigate crops and the contamination of soils by e-waste are often cited as the primary sources of pollution in most, if not all, of these cases. According to scientific literature and electronic media sources, most of the contamination is in China's top rice-growing province of Hunan, which is reported to have more than 200 mills that produce about 2 million tons of rice per year (97). Annual harvest of rice in the province is reported to exceed 26 million tons, accounting for 13 percent of the country's rice production (97). The principal toxic heavy metals contaminants that are frequently reported are As, Pb and Cd. These findings could as well be just the tip of the iceberg and symptomatic of a nationwide crisis. According to Zitan (97), one-third of China's rice contains high levels of Pb, and one-tenth contains high levels of Cd. Another article reported that 60 percent of rice in Jiangxi, Hunan and Guangdong provinces contained elevated Cd (100).

The most disturbing development was that after the discovery that most rice in the Hunan province of China was contaminated with unsafe levels of heavy metals, the rice was not destroyed but was sold in another province at a lower price. Some people in China do not blame the "hardworking farmers" but the government which they say "does not protect the ecosystems, it allows industrial waste to be released onto farmland, and it is poisoning the people." (97). According to Zitan (97), the Ministries of Environmental Protection and Land and Resources had data that showed the magnitude of the nationwide soil pollution back in 2006, but did not make the data public, instead declared the results a "state secret". It only became public after a leaked document showed that 89,000 acres of farmland in China were severely contaminated with heavy metals, potentially affecting 12 million tons of crops annually (97).

Thailand, a country that contributes a significant amount of rice to the world population, has its agricultural soils reportedly contaminated by As, Cd, Pb and other heavy metal(loids) from industrial and mining activities, solid waste management and agro-chemicals (100–102). Literature has it that irrigation water sources in the country are also contaminated with toxic heavy metal(loids) (103, 104). Other studies have also shown that some agricultural products in the country contain elevated levels of Pb. For instance, a study by Sang-Uthai, et al. (105) reported Pb levels in garlic and peppercorns as high as 10.908 mg/kg and 10.484 mg/kg respectively.

In the Czech Republic, particularly in the Pribram region, agricultural soils are reported to be heavily polluted by atmospheric deposition from lead-smelters (106). The concentration of Pb in the agricultural soils in this region was reported to be as high as 2500 mg/kg. Other significant anthropogenic sources of contamination of agricultural soils in the Czech Republic include industrial activities and soil acidification (100, 101, 107). Researchers in the country have expressed concerns regarding the continued accumulation of metals in agricultural soils, industrial and mining activities, and its potential impact on human health.

It has to be pointed out that most regions of the world are contaminated by environmental pollutants from anthropogenic activities which may vary in scope from country to country. For example, a comprehensive analysis of As in rice and rice food products by the FDA showed some rice samples from the USA with higher levels of As than rice samples from India and Thailand (14). This shows that each region has its own problems of environmental pollution by specific pollutants and, therefore, a role to play in protecting human health from food contamination. In other countries, what is lacking is research data about food contamination in the public domain, not environmental and food contamination. Anthropogenic activities worldwide suggest a potential global crisis.

Pragmatic Approaches to Food Contamination

Food Testing and Safe Limits

There is a substantial amount of data in the literature on the levels of environmental pollutants in food. However, the data cannot be compiled into a public database for the purposes of comprehensively assessing the human health risk that is posed by food consumption. This is primarily due to lack of standard protocols and analytical measurement techniques that include acceptable quality assurance or quality control, and enforceable safe limits of environmental pollutants in food.

In the literature involving food contamination, study objectives, measurement methods and quality control methods vary. Objectives usually set the tone for: the entire research activity; type and number of samples; sampling strategy; kind of data; measurement techniques; and use of data. Standard protocols and analytical measurement techniques that include acceptable quality assurance or quality control must be developed and implemented to make research data on food contamination comparable across laboratories and time periods.

In addition to standard protocols and analytical measurement techniques, safe limits of environmental pollutants in food have to be established to put meaning, relevance, and significance to data from scientific research. Currently there are numerous guidelines that are enforceable and yet subject to manipulation, as well as subjective interpretation. Conducting research that has a bearing on public health, such as determining concentrations of chemical pollutants in food without established safe limits to which the experimental data can be referenced or compared has no scientific merit and will not serve the intended purpose well. Such data confuses the consumer and puts human health at risk. For instance, many researchers have published data on the levels of As in rice and rice food products, and fruit juices (*108–115*) in which As levels in a significant number of samples were reported to exceed recommended guidelines. The findings may not be of any significance or consequence though, because the guidelines are not enforced and people may still be consuming the foods.

The FDA has made tremendous efforts in conducting comprehensive testing of toxic metals in foods and maintaining a database of the results on their website. However, having the data with safe limits to reference the data may result in subjective interpretations and mixed messages that may put human health at risk. In September 2013, the FDA released comprehensive As test results for more than 1,300 rice and rice food products samples (*14*). Inorganic As (iAs) not total As (tAs) in rice samples ranged from 3.5 to 7.2 μg/serving; infant formula 0.1 μg/serving (148 g serving); infant cereal 1.8 μg/serving; toddler cereal 1.5 μg/serving. The levels of tAs were greater than the iAs levels.

The FDA's conclusion was, "*The levels FDA found in its testing are too low to cause immediate or short-term adverse health effects. FDA's work going forward will center on long-term risk and ways to manage it with a focus on long-term exposure.*" Again, in the absence of established safe limits, such a conclusion may put human health at risk, especially given that the levels of As in a large number of the beverages tested were higher than the established safe limit of 10 ppb in drinking water. Also, they may have used conservative estimates to calculate the amount per serving; people tend to eat more rice and more frequently than what some may consider normal. In some ethnic groups, rice and rice products form part of their diet almost on a daily basis, and they obviously eat more than other ethnic groups. Research has also shown that environmental pollutants, including As, do not have to be in large amounts to cause adverse health effects (20, 26). Low levels that may be deemed harmless may have irreversible chronic effects to humans if exposed during the early stages of development (7). There is also the effect of simultaneous exposures from multiple sources that may include the rice.

Ultimately, enforceable safe limits that take into account multiple exposures need to be established. Safe limits must be "interactive". Consumers in locations prone to simultaneous exposures from multiple sources should have safe limits that set at lower thresholds in individual sources than consumers who are less vulnerable to multiple exposures. To its credit, the FDA is reportedly working on the risk assessment of As exposure which they expect to release in 2014, after which decisions on proposed voluntary or mandatory limits of As in rice and rice products or other steps as necessary will be made. In the meanwhile, the Agency is currently advising consumers, including pregnant women and children, to avoid consuming an excess of any one food by eating a well-balanced diet. The recommendation makes sense provided the other foods are not also contaminated.

Most studies have focused on As in foods, particularly in rice and rice food products as well as in fruits and fruit juices and vegetables, primarily because these foods have been identified as the primary source of the metalloid (13). In fact, the FDA has been testing for total As in a variety of foods that include rice and juices since 1991 through its Total Diet Study program (14). According to its website, the agency also monitors toxic elements, including As in selected domestic and imported food, under the Toxic Elements Program, including children's food and beverages (116). Other toxic heavy metals that have been on the FDA's radar are Hg and Pb. Mercury is only tested in fish (117) according to the data available on the Agency's website. With regard to Pb, the Agency regularly tests for the heavy metal in foods and beverages that are commonly consumed by children through its Total Diet Study (118). Children are one of the population subgroups that are vulnerable to Pb toxicity. However, the elderly, and pregnant or lactating women, are also vulnerable to the effects of lead toxicity. The selective testing, though largely understandable, may endanger other vulnerable population sub-groups.

In Europe, the European Food Safety Authority (EFSA) is also doing a sterling job in its quest to meet its regulatory obligations to the European Commission and European Union Member States (119). Just like the FDA in the United States, EFSA is actively involved in regulatory issues relating to food contamination. As is the case with the FDA, the focus regarding metals and metalloids in food is on toxic, non-essential elements, particularly As, Cd, Pb, and Hg, with one of the major objectives being to establish harmonized safe limits for the toxic metals and metalloids in food. On the October 22nd, 2009, EFSA's panel on contaminants in the food chain (CONTAM Panel) recommended the reduction in exposure to inorganic arsenic via food, and emphasized the need for more speciation data (organic and inorganic) for the metalloid in food as well as potential health effects related to different levels of intake (119). The CONTAM Panel identified cereal grains and cereal-based products, food for special dietary uses (such as algae), bottled water, coffee and beer, rice and rice-based products, fish and vegetables as the primary dietary sources of inorganic arsenic (119). As a result of the absence of harmonized safety limits for As in foodstuffs in Europe, the European Commission asked EFSA to assess the health risks related to the presence of arsenic as a contaminant in foodstuffs (119).

With regard to Cd, EFSA's CONTAM Panel in 2009 established a tolerable weekly intake (TWI) of 2.5 µg/kg body weight which they reaffirmed in 2011 following a Joint FAO/WHO Expert Committee on Food Additives (119). The Panel noted then that the current average dietary exposure to Cd for adults was close to the TWI and that the exposure of subgroups, such as children, vegetarians and people living in highly contaminated areas, at that time, could exceed the TWI (119). As a result, the Panel recommended reduction in Cd exposure at the population level (119). The Panel, in its opinion published in April 2010, also raised concerns on the possible adverse health effects in fetuses, infants and children, even though they had concluded that current levels of exposure to lead then did not pose a significant risk for most adults (119). Cereals, vegetables and tap water were identified as the significant contributors to dietary exposure to Pb for most Europeans (119).

Given the extent of pollution of agricultural soils and food crops across the globe as reported in the literature, the vulnerability of food to chemical contamination, and the globalized food market, all foods have to be regularly screened for both non-essential toxic elements and essential toxic elements against established safe standards, and all data generated placed in a database that is accessible to all. Consumers may be greatly helped by having levels of potential toxicants, even if lower than established safe limits, put labels on food packaging to avoid inadvertent toxicity that may result from simultaneous exposure from more than one food source. The trend in pollution must, however, be curtailed for a lasting solution to the problem of food contamination.

Prevention May Be the Key

Comprehensive testing of all foods can be done, and mandatory safe limits established, but if the trend in food contamination continues, the mandatory safe limits will be surpassed and then what? Are people expected to put resources into the production of food that they know may be contaminated and throw it away in the event that the mandatory limits are surpassed upon testing? Consumers cannot be made to choose between the possibility of dying from hunger or dying from the debilitating chronic health effects of environmental toxicants. The best solution may be prevention which has long been known to be better than a cure; only that in this case there may not be any cure because contaminated food may not be remedied. Either it is consumed it or dumped. Measures have to be put in place to prevent food contamination, and focus has to be on the source and not the product.

Curtailing pollution in modern times is a difficult proposition. As a result, wild thoughts can be pardoned. What about the following two-pronged approach: remediate contaminated agricultural soils and irrigation waters, and mitigate to prevent new pollution while developing high-yield crop varieties that will take up less toxic chemical pollutants from soil and water? Granted; this is an ambitious and costly measure, but considering what is at stake, it may be worth the consideration. In any case, a large amount of resources are already being expended in testing and screening foods for contaminants. These resources could be used to address the problem at the source.

Engaging in unethical agricultural practices such as the use of sewage and industrial effluents that are known to be contaminated to irrigate crops, and the use of contaminated bio-solids (sewage sludge) as fertilizer or soil amendment must be stopped by national and international laws. We live in a global village so we should be able to chastise wayward fellow global citizens, especially when their activities have bearings on the wellbeing of the rest. As well, developed nations must stop the shipments of harmful waste such as e-waste to poor nations, regardless of the purpose. We need environmental justice in the global village.

Water shortages in the developing countries can be remedied by developing infrastructure such as dams to harness and store rainwater, and infrastructure such as boreholes to utilize groundwater. In Africa for instance, groundwater is considered the major source of drinking water on the continent, yet it remains under-utilized. In fact, the use of groundwater in Africa to irrigate crops is projected to increase considerably to combat growing food shortages (*120*). The assumption is that the required resources to utilize the groundwater will be available and widely accessible. This could be a figment of imagination by all accounts. They will need help.

Because of the enormous amount of resources that will be required for remediating contaminated soils and waters, rich nations will have to assist poor nations in their decontamination efforts in exchange for food or other forms of payment. This will benefit both the poor and the rich in the long term. Agriculture must be practiced away from the influence of contaminating sources such as industries, mines, and waste management sites. With high yield varieties, enough food can be produced in relatively less acreage than would be required with low yield varieties. With adequate agricultural infrastructure, and climate and weather patterns permitting, agriculture can be practiced throughout the year. Adequate food will be produced.

Crafting and policing international laws is not an enviable task, but we may be heading in that direction. The starting point may be easy. There is an existing international body that deals with food; the Food and Agriculture Organization (FAO), an agency of the United Nations that deals with international efforts to prevent hunger in both the developed and developing countries. The agency can be empowered and mandated to ensure food safety in addition to its current mandate of ensuring food security for all nations. In any case, agricultural infrastructure modernization in member states of the United Nations is already under the auspices of the FAO. FAO will be responsible for the funding, implementation and supervision of the decontamination efforts by nations as well as legally holding nations accountable should they engage in activities that are contrary to the laid down statutes. With sufficient resources, the FAO can also keep a database of regular soil, water, and food testing results of all nations that is accessible to any interested party. Local agencies in individual nations such as the USDA and FDA in the US will be required to work closely with the international agency.

References

1. United Nations Department of Economics and Social Affairs Population Division: 24/2/2005 Press Release. http://www.un.org/News/Press/docs/2005/pop918.doc.htm (accessed October 22, 2013).
2. World Population Growth. http://www.worldbank.org/depweb/english/beyond/beyondco/beg_03.pdf (accessed October 22, 2013).
3. Engelman, R.; et al. People in the Balance. Washington, DC: Popular Action International, 2000. *Nature* **2003**, *422*, 252 (as reproduced in *Nature*).
4. Unites States Department of Agriculture, Economic Research Service: U.S. Food Imports. http://www.ers.usda.gov/data-products/us-food-imports.aspx (accessed October 22, 2013).
5. Wang, X.; Nan, Z.; Liao, Q.; Ding, W.; Wu, W. *Pol. J. Environ. Stud.* **2012**, *21*, 1867–1874.
6. Murtaza, G.; Ghafoor, A.; Qadir, M.; Owens, G.; Aziz, M. A.; Zia, M. H.; Saifullah *Pedosphere* **2010**, *20*, 23–34.
7. National Institute of Environmental Health Sciences: Child Development and Environmental Toxins. https://www.niehs.nih.gov/health/assets/docs_a_e/child_development_and_environmental_toxins_508.pdf (accessed October 22, 2013).
8. Koturbash, I.; Beland, F. A.; Pogribny, I. P. *Toxicol. Mech. Methods* **2011**, *21*, 289–297.
9. Thayer, K. A.; Heindel, J. J.; Bucher, J. R.; Gallo, M. A. *Environ. Health Perspect.* **2012**, *120*, 779–789.
10. Fragou, D.; Fragou, A.; Kouidou, S.; Njau, S.; Kovatsi, L. *Toxicol. Mech. Methods* **2011**, *21*, 343–352.
11. Jackson, B. P.; Taylor, V. F.; Karagas, M. R.; Punshon, T.; Cottingham, K. L. *Environ. Health Perspect.* **2012**, *120*, 623–626.
12. U.S. Food and Drug Administration: Arsenic in Rice and Rice Products. http://www.fda.gov/Food/FoodborneIllnessContaminants/Metals/ucm319870.htm (accessed October 22, 2013).
13. Davis, M. A.; Mackenzie, T. A.; Cottingham, K. L.; Gilbert-Diamond, D.; Punshon, T.; Karagas, M. R. *Environ. Health Perspect.* **2012**, *120*, 1418–24.
14. U.S. Food and Drug Administration: Arsenic. http://www.fda.gov/Food/FoodborneIllnessContaminants/Metals/ucm280202.htm (accessed October 22, 2013).
15. Carey, A.; Scheckel, K. G.; Lombi, E.; Newville, M.; Choi, Y.; Norton, G. J.; Charnock, J. M.; Feldmann, J.; Price, A. H.; Meharg, A. A. *Plant Physiol.* **2010**, *152*, 309–319.
16. Heitkemper, D. T.; Kubachka, K. M.; Halpin, P. R.; Allen, M. N.; Shockey, N. V. *Food Addit. Contam., Part B* **2009**, *2*, 112–120.
17. Tuli, R.; Chakrabarty, D.; Trivedi, P. K.; Tripathi, R. D. *Mol. Breed.* **2010**, *26*, 307–323.
18. Ren, X.; McHale, C. M.; Skibola, C. F.; Smith, A. H.; Smith, M. T.; Zhang, L. *Environ. Health Perspect.* **2011**, *119*, 11–19.
19. National Institute of Environmental Health Sciences: Environmental Factor, July 2011. Superfund postdoc unravels arsenic exposure. http://

www.niehs.nih.gov/news/newsletter/2011/july/science-superfund/index.cfm (accessed October 22, 2013).

20. Cronican, A. A.; Fitz, N. F.; Carter, A; Saleem, M; Shiva, S; et al. *PLoS ONE* **2013**, *8*, e53478.

21. Farzan, S. F.; Karagas, M. R.; Chen, Y. *Toxicol. Appl. Pharmacol.* **2013**, *272*, 384–390.

22. Fei, D. L.; Koestler, D. C; Li, Z.; Giambelli, C.; Sanchez-Mejias, A.; Gosse, J. A.; Marsit, C. J; Karagas, M. R; Robbins, D. J. *Environ. Health: Global Access Sci. Source* **2013**, *12*, 58.

23. Moon, K.; Guallar, E.; Navas-Acien, A. *Curr. Atheroscler. Rep.* **2012**, *14*, 542–555.

24. Liaw, J; et al. *Cancer Epidemiol., Biomarkers Prev.* **2008**, *17*, 1982–1987.

25. Yorifuji, T; et al. *Environ. Health Prev. Med.* **2011**, *16*, 164–170.

26. Pilsner, J. R.; Hall, M. N.; Liu, X.; Ilievski, V.; Slavkovich, V.; et al. Influence of Prenatal Arsenic Exposure and Newborn Sex on Global Methylation of Cord Blood DNA. *PLoS ONE* **2012**, *7*, e37147.

27. Marsit, C. J.; Karagas, M. R.; Robbins, D. J. *Environ. Health: Global Access Sci. Source* **2013**, *12*, 58.

28. National Institute of Environmental Health Sciences: Health Homes = Health Kids. http://www.niehs.nih.gov/health/assets/docs_f_o/health-kids-water.pdf (accessed October 22, 2013).

29. Riederer, A. M.; Belova, A.; George, B. J.; Anastas, P. T. *Environ. Sci. Technol.* **2013**, *47*, 1137–1147.

30. Satarug, S.; Moore, M. R. *Tohoku J. Exp. Med.* **2012**, *228*, 267–288.

31. Ciesielski, T.; Bellinger, D. C.; Schwartz, J.; Hauser, R.; Wright, R. O. *Environ. Health* **2013**, *12*, 13.

32. Ciesielski, T.; Weuve, J.; Bellinger, D. C.; Schwartz, J.; Lanphear, B.; Wright, R. O. *Environ. Health Perspect.* **2012**, *120*, 758–763.

33. Zhang, L.; Lv, J.; Liao, C. *Biol. Trace Element Res.* **2012**, *146*, 287–292.

34. Meharg, A. A.; Norton, G.; Deacon, C.; Williams, P.; Adomako, E. E.; Price, A.; Zhu, Y.; Li, G.; Zhao, F-J.; McGrath, S.; Villada, A.; Sommella, A.; De Silva, P. M. C. S.; Brammer, H.; Dasgupta, T.; Islam, M. R. *Environ. Sci. Technol.* **2013**, *47*, 5613–5618.

35. De Silva, P. M. C. S.; Brammer, H.; Dasgupta, T.; Islam, M. R. *Environ. Sci. Technol.* **2013**, *47*, 5613–5618.

36. He, P.; Lu, Y.; Liang, Y.; Chen, B.; Wu, M.; Li, S.; He, G.; Jin, T. *BMC Public Health* **2013**, *13*, 590.

37. Cory-Slechta, D. A; et al. *Basic Clin. Pharmacol. Toxicol.* **2008**, *102*, 218–227.

38. National Institute of Environmental Health Sciences: Kids pages, Lead Poisoning. http://kids.niehs.nih.gov/explore/pollute/lead.htm (accessed October 22, 2013).

39. *Technology Planning and Management Corporation. Report on Carcinogens background Document for Lead and Lead Compounds*; Contract Number N01-ES-85421; U.S. Department of Health and Human Services: Durham, NC, May 8, 2003. http://ntp.niehs.nih.gov/ntp/newhomeroc/roc11/Lead-Public.pdf#search=lead (accessed October 2013).

40. National Institute of Environmental Health Sciences: Lead. http://www.niehs.nih.gov/health/assets/docs_f_o/lead-fs.pdf (accessed October 22, 2013).

41. Wang, X.; Nan, Z.; Liao, Q.; Ding, W.; Wu, W. *Pol. J. Environ. Stud.* **2012**, *21*, 1867–1874.

42. Murtaza, G.; Ghafoor, A.; Qadir, M.; Owens, G.; Aziz, M. A.; Zia, M. H.; Saifullah *Pedosphere* **2010**, *20*, 23–34.

43. Neal, A. P.; Guilarte, T. R. *Toxicol. Res.* **2013**, *2*, 99–114.

44. *Lenntech Water Treatment Solutions: Health Effects of Manganese.* http://www.lenntech.com/periodic/elements/mn.htm#ixzz2CmFJZstt (accessed October 22, 2013).

45. National Institutes of Health: Office of Dietary Supplements. http://ods.od.nih.gov/factsheets/Zinc-HealthProfessional/ (accessed October 22, 2013).

46. Carbonell-Barrachina, A. A.; Ramírez-Gandolfo, A.; Wu, X.; Gareth, J.; Norton, G. J.; Burló, F.; Deacon, C.; Meharg, A. A. *J. Environ. Monit.* **2012**, *14*, 2447–2455.

47. Nookabkaew, S.; Rangkadilok, N.; Mahidol, C.; Promsuk, G.; Satayavivad, J. *J. Agric. Food Chem.* **2013**, *61*, 6991–6998.

48. Bennett, J. P.; Chiriboga, E.; Coleman, J.; Waller, D. M. *Sci. Total Environ.* **2000**, *246*, 261–9.

49. Tongesayi, T.; Fedick, P.; Lechner, L.; Brock, C.; Le Beau, A.; Bray, C. *Food Chem. Toxicol.* **2013**, *62*, 142–147.

50. Osibanjo, O.; Nnorom, I. C.; Bakare, A. A.; Alabi, O. A. *Adv. Environ. Res.* **2012**, *17*, 403–433.

51. Samanta, T. *Everyman's Sci.* **2010**, *45* (1), 44–49.

52. Pia, T. *Acta Material.* **2013**, *61*, 1001–1011.

53. Robinson, B. H. *Sci. Total Environ.* **2009**, *408*, 183–191.

54. Ladou, J.; Lovegrove, S. *Int. J. Toxicol., Occup. Environ. Health* **2008**, *14*, 1–10.

55. Sthiannopkao, S.; Wong, M. H. *Sci. Total Environ.* **2013**, *463-464*, 1147–1153.

56. Anon, A. *Environ. Sci. Technol.* **2002**, *36*, 309A–310A.

57. Chen, A.; Dietrich, K. N.; Huo, X.; Ho, S. *Environ. Health Perspect.* **2011**, *119*, 431–8.

58. Gupta, R.; Sangita; Kaur, V. *Res. J. Chem. Sci.* **2011**, *1*, 49–56.

59. Tang, X.; Shen, C.; Chen, L.; Xiao, X.; Wu, J.; Khan, M. I.; Dou, C.; Chen, Y. *J. Soils Sediments* **2010**, *10*, 895–906.

60. Fujimori, T.; Takigami, H.; Agusa, T.; Eguchi, A.; Bekki, K.; Yoshida, A.; Terazono, A.; Ballesteros, F. C., Jr. *J. Hazardous Mater.* **2012**, *221–222*, 139–146.

61. Ha, N. N.; Agusa, T.; Ramu, K.; Tu, N. P. C.; Murata, S.; Bulbule, K. A.; Parthasaraty, P.; Takahashi, S.; Subramanian, A.; Tanabe, S. *Chemosphere* **2009**, *76*, 9–15.

62. Luo, C.; Liu, C.; Wang, Y.; Liu, X.; Li, F.; Zhang, G.; Li, X. *J. Hazardous Mater.* **2011**, *186*, 481–490.

63. Fu, J.; Zhou, Q.; Liu, J.; Liu, W.; Wang, T.; Zhang, Q.; Jiang, G. *Chemosphere* **2008**, *71*, 1269–1275.

64. Mueller, B.; Scheytt, T.; Asbrand, M.; de Casas, A. M. *Hydrogeol. J.* **2012**, *20*, 1117–1129.

65. Al Omron, A. M.; El-Maghraby, S. E.; Nadeem, M. E. A.; El-Eter, A. M.; Al-Mohani, H. *J. Saudi Soc. Agric. Sci.* **2012**, *11*, 15–18.

66. El-Motaium, R. A. Environmental risk assessment of prolonged use of sewage wastewater in irrigation at El-Gabal El-Asfar farm: alleviation using ionizing radiation. Presented at Cairo International Conference on Energy and Environment, 11th, Hurghada, Egypt, Mar. 15−18, 2009, 12/1−12/11.

67. Lottermoser, B. G. *Environ. Geochem. Health* **2012**, *34*, 67–76.

68. New Jersey Department of Environmental Protection, Solid and Hazardous Waste, State Wide Solid Waste Management Plan 2006 Section K - Statewide Sludge Management Plan pp K1-K45. http://www.state.nj.us/dep/dshw/recycling/swmp/pdf/section_k_06.pdf (accessed October 22, 2013).

69. Tiwari, K. K.; Singh, N. K.; Patel, M. P.; Tiwari, M. R.; Rai, U. N. *Ecotoxicol. Environ. Saf.* **2011**, *74*, 1670–1677.

70. Rana, L.; Dhankhar, R.; Chhikara, S. *Int. J. Environ. Res.* **2010**, *4*, 513–518.

71. Sharma, R. K.; Agrawal, M.; Marshall, F. *Ecotoxicol. Environ. Saf.* **2007**, *66*, 258–266.

72. Patel, K. P.; Singh, M. V.; Ramani, V. P.; Patel, K. C.; George, V.; Zizala, V. J. *Pollut. Res.* **2006**, *25*, 25–30.

73. Khurana, M. P. S.; Singh, M.; Nayyar, V. K. *Indian J. Environ. Ecoplann.* **2004**, *8*, 221–228.

74. Karthikeyan, K.; Singh, K. *J. Ind. Pollut. Control* **2004**, *20*, 211–219.

75. Madhavi, A.; Rao, A. P.; Ramavatharam, N. *Pollut. Res.* **2004**, *23*, 33–437.

76. Patel, K. P.; Pandya, R. R.; Maliwal, G. L.; Patel, K. C.; Ramani, V. P.; George, V. *J. Indian Soc. Soil Sci.* **2004**, *52*, 89–94.

77. Rawat, M.; Ramanathan, Al.; Subramanian, V. *J. Hazardous Mater.* **2009**, *172*, 1145–1149.

78. Sherene, T. *J. Ecotoxicol. Environ. Monit.* **2009**, *19*, 397–400.

79. Prabhavathi, M.; Reddy, K. S.; Indirani, R. *Ind. J. Environ. Ecoplan.* **2010**, *17*, 343–350.

80. Singh, J.; Upadhyay, S. K.; Pathak, R. K.; Gupta, V. *Toxicol. Environ. Chem.* **2011**, *93*, 462–473.

81. Raj, G. B.; Patnaik, M. C.; Babu, P. S.; Kalakumar, B.; Singh, M. V.; Shylaja, J. *Ind. J. Animal Sci.* **2006**, *76*, 131–133.

82. Huang, G.-X.; Chen, Z.-Y.; Sun, J.-C.; Liu, J.-T.; Zhang, Y.-X.; Wang, J.-C. *J. Central South University* **2012**, *19*, 2620–2626.

83. Wang, Y.; Zhang, P.; Chen, Y. *Anhui Nongye Kexue* **2012**, *40*, 8512–8514.

84. Wang, X.; Nan, Z.; Liao, Q.; Ding, W.; Wu, W. *Pol. J. Environ. Stud.* **2012**, *21*, 1867–1874.

85. Chen, Z. S. *Biogeochem. Trace Met.* **1992**, 85–108.

86. Wong, S. C.; Li, X. D.; Zhang, G.; Qi, S. H.; Min, Y. S. *Environ. Pollut.* **2002**, *119*, 33–44.

87. Feng, J.; Zhao, J.; Bian, X.; Zhang, W. *Environ. Geochem. Health* **2012**, *34* (5), 605–614.

88. Huang, Z.-Y.; Chen, T.; Yu, J.; Qin, D.-P.; Chen, L. *Environ. Geochem. Health* **2012**, *34* (1), 55–65.

89. Li, J. X.; Yang, X. E.; He, Z. L.; Jilani, G.; Sun, C. Y.; Chen, S. M. *Geoderma* **2007**, *141*, 174–180.

90. Hseu, Z.-Y.; Su, S.-W; Lai, H.-Y.; Guo, H.-Y.; Chen, T.-C.; Chen, Z.-S. *Soil Sci. Plant Nutrit.* **2010**, *56*, 31–52.

91. Lai, H.-Y.; Hseu, Z.-Y.; Chen, T.-C.; Chen, B.-C.; Guo, H.-Y.; Chen, Z.-S. *Int. J. Environ. Res. Public Health* **2010**, *7*, 3595–3614.

92. Staff Reporter. Harmful heavy metals found in Hunan-produced rice. *China Times* 2013-03-01. http://www.wantchinatimes.com/news-subclass-cnt.aspx?cid=1202&MainCatID=12&id=20130301000004 (accessed October 22, 2013)

93. Jing, G. Heavy Metals Tainting China's Rice Bowls. *Caixin Online* 2011-02-13. http://english.caixin.com/2011-02-13/100224762.html (accessed October 22, 2013).

94. China rice laced with heavy metals: report. *PHYS.ORG* 2011-02-16. http://phys.org/news/2011-02-china-rice-laced-heavy-metals.html (accessed October 22, 2013).

95. Xinzhen, L. The Crisis of Tainted Rice: Soil pollution is impacting one of China's long-cherished staples. What's to be done? *China's National English News Weekly, BeijingReview.COM* 2013-04-11. http://www.bjreview.com/quotes/txt/2013-04/11/content_533012.htm (accessed October 22, 2013).

96. Associated Press. New Rice Contamination Reported in China. *EPOCH TIMES* 2013-05-13. http://www.theepochtimes.com/n3/67127-new-rice-contamination-reported-in-china/ (accessed October 22, 2013).

97. Zitan, G. Heavy Metal Pollution Threatens China's Rice Industry. *Epoch Times* 2013-04-03. http://www.theepochtimes.com/n3/8839-heavy-metal-pollution-threatens-chinas-rice-industry/ (accessed October 22, 2013).

98. Heavy metal fears devastate market for rice farmers. *Sina English* 2013-03-26. http://english.sina.com/china/2013/0325/575294.html (accessed October 22, 2013).

99. Shahane, N. China Grown Rice Defiled With Heavy Metals and Chemicals. *TOPNEWS* 2011-02-16 .http://topnews.net.nz/content/211961-china-grown-rice-defiled-heavy-metals-and-chemicals (accessed October 22, 2013).

100. Chuangcham, U.; Wirojanagud, W.; Charusiri, P.; Milne-Home, W.; Lertsirivorakul, R. *J. Appl. Sci.* **2008**, *8*, 1383–1394.

101. Prechthai, T.; Parkpian, P.; Visvanathan, C. *J. Hazardous Mater.* **2008**, *156*, 86–94.

102. Panichayapichet, P.; Nitisoravut, S.; Simachaya, W.; Wangkiat, A. *Water, Air, Soil Pollut.* **2008**, *194*, 259–273.

103. Chotpantarat, S.; Ong, S. K.; Sutthirat, C.; Osathaphan, K. *J. Sci. Res. Chulalongkorn University* **2008**, *33*, 101–110.

104. Tupwongse, V.; Parkpian, P.; Watcharasit, P.; Satayavivad, J. *J. Environ. Sci. Health, Part A: Toxic/Hazard. Subst. Environ. Eng.* **2007**, *42*, 1029–1041.

105. Sang-Uthai, K.; Charoenteeraboon, J.; Phaechamud, T.; Limmatvapirat, C. Heavy metal contamination of red and green curry paste in some areas of Thailand. In *Kamphaengsaen International Natural Products Symposium:*

The Relationship between Living Organisms and Environment, Proceeding Book, 1st, Bangkok, Thailand, Oct. 23−24, 2010; pp 122−128.

106. Sichorova, K.; Tlustos, P.; Szakova, J.; Korinek, K.; Balik, J. *Plant, Soil Environ.* **2004**, *50*, 525–534.

107. Rautengarten, A. M.; Schnoor, J. L.; Anderberg, S.; Olendrzynski, K.; Stigliani, W. M. *Water, Air, Soil Pollut.* **1995**, *85*, 737–42.

108. Carey, A.; Scheckel, K. G.; Lombi, E.; Newville, M.; Choi, Y.; Norton, G. J.; Charnock, J. M.; Feldmann, J.; Price, A. H.; Meharg, A. A. *Plant Physiol.* **2010**, *152*, 309–319.

109. Heitkemper, D. T.; Kubachka, K. M.; Halpin, P. R.; Allen, M. N.; Shockey, N. V. *Food Addit. Contam., Part B* **2009**, *2*, 112–120.

110. Tuli, R.; Chakrabarty, D.; Trivedi, P. K.; Tripathi, R. D. *Mol. Breed.* **2010**, *26*, 307–323.

111. Roberge, J.; Abalos, A. T.; Skinner, J. M.; Kopplin, M.; Harris, R. B. *Am. J. Environ. Sci.* **2009**, *5*, 688–694.

112. Gilbert-Diamond, D.; Cottingham, K. L.; Gruber, J. F.; Tracy Punshon, T.; Sayarath, V.; Gandolfi, A. J.; Baker, E. R.; Jackson, B. P.; Folt, C. L.; Karagas, M. R. *Proc. Natl. Acad. Sci. U.S.A.* **2011**, *108*, 20656–20660.

113. Huang, J.-H.; Fecher, P.; Ilgen, G.; Hu, K.-N.; Yang, J. *Food Chem.* **2012**, *130*, 453–459.

114. Bhattacharya, P.; Samal, A. C.; Majumdar, J.; Santra, S. C. *Water, Air Soil Pollut.* **2010**, *213*, 3–13.

115. Narukawa, T.; Chiba, K. *J. Agric. Food Chem.* **2010**, *58*, 8183–8188.

116. U.S. Food and Drug Administration. Questions & Answers: Arsenic in Rice and Rice Products. http://www.fda.gov/Food/FoodborneIllnessContaminants/Metals/ucm319948.htm (accessed October 22, 2013).

117. U.S. Food and Drug Administration. Mercury and Methylmercury. http://www.fda.gov/Food/FoodborneIllnessContaminants/Metals/ucm088758.htm 2009 (accessed October 22, 2013).

118. U.S. Food and Drug Administration. Reported Findings of Low Levels of Lead in Some Food Products Commonly Consumed by Children. http://www.fda.gov/Food/FoodborneIllnessContaminants/Metals/ucm233520.htm (accessed October 22, 2013).

119. The European Food Safety Authority: Metals as contaminants in food. http://www.efsa.europa.eu/en/topics/topic/metals.htm?wtrl=01 (accessed January 24, 2014)

120. MacDonald, A. M.; Bonsor, H. C.; Dochartaigh, B. E. O.; Taylor, R. G. Quantitative maps of groundwater resources in Africa. *Environ. Res. Lett.* **2012**, *7*, 024009.

Chapter 4

FDA's Regulation of Nanotechnology in Food Ingredients

Teresa A. Croce*

Division of Petition Review, Office of Food Additive Safety, U.S. Food and Drug Administration, 5100 Paint Branch Parkway, HFS-265, College Park, Maryland 20740
*E-mail: Teresa.Croce@fda.hhs.gov

Nanotechnology has the potential to impact many FDA-regulated product areas, including food ingredients. The agency issued draft guidance in 2011 to address its current thinking on nanotechnology and has extended that overarching guidance through the publication of a subsequent draft guidance document in 2012 specific to food ingredients, including food contact substances and food ingredients that are color additives. The 2012 document examines how significant changes in manufacturing process, including the integration of nanotechnology, may impact the safety and regulatory status of the food ingredient.

Introduction

The U.S. Food and Drug Administration (FDA) is responsible for ensuring that food ingredients meet safety and quality standards mandated by U.S. law. In doing so, the agency bases its regulatory decisions on the best available scientific data and information (*1*). Ensuring that food ingredients are safe while fostering innovation of new products requires that FDA maintain its scientific expertise and stays current with evolving science and technology. New technologies, such as nanotechnology, present unique opportunities for innovation along with potential challenges for the regulatory authorities tasked with protecting consumer and environmental health. FDA's regulatory authority covers a wide range of product areas, from evaluation of the safety and efficacy of drugs and devices to the safety of the food supply (*2*). Many FDA-regulated product areas could be impacted by advances in basic and applied nanotechnology (*4*). Different statutory authorities govern how FDA regulates the various product areas. While FDA is routinely faced with assessing the risk associated with new technologies, the mechanism to deal with those uncertainties may differ due to these governing statutory requirements.

The Center for Food Safety and Applied Nutrition (CFSAN) is one of six product Centers located within FDA and is responsible for protecting and promoting the public health by ensuring that the Nation's food supply is safe, sanitary, wholesome, and properly labeled, as well as for ensuring that cosmetic products are safe and properly labeled (*5*). Nanotechnology applications are relevant to many of CFSAN's product areas, including foods and cosmetics. CFSAN's Office of Food Additive Safety is responsible for ensuring the safety of substances added to food, including food additives, color additives, food contact substances, and generally recognized as safe (GRAS) ingredients.

Regulation of Food Ingredients in the United States

Food Additives and Color Additives

FDA's authority to regulate food additives was established in 1958 with the passage of the Food Additives Amendment to the Federal Food, Drug, and Cosmetic Act (FD&C Act). This amendment required that food additives undergo premarket evaluation to establish that their intended use is safe. Specifically, section 201(s) of the FD&C Act defined the term "food additive" (*6*) to include both direct and indirect additives and provided an exemption for ingredients that are GRAS (*7*). Section 409 of the FD&C Act laid out the general safety standard for food additives and GRAS substances and set forth the requirements that food additives must undergo premarket approval by FDA. In 1960, the Color Additive Amendments were enacted which defined "color additive" (*8*) and required that only color additives listed as "suitable and safe" are used in foods, drugs, and cosmetics.

Following the passage of these amendments, FDA updated its regulations to establish a petition process, which is the mechanism for seeking premarket approval of food additives and color additives. These requirements are found in Title 21 of the Code of Federal Regulations (CFR) Parts 170 and 171 for food additives and Parts 70 and 71 for color additives. An interested party can prepare and submit a food additive or color additive petition for either a new or an expanded use of an additive. Importantly, the burden lies with the petitioner to demonstrate safety by submitting the necessary data and information. FDA performs a fair evaluation of the data submitted and makes a determination of safety. If the petitioner has demonstrated to FDA's satisfaction that there is a reasonable certainty of no harm from the intended use of the additive, a regulation will publish in 21 CFR establishing safe conditions of use for the additive. It is important to note that foods containing an unapproved food additive may not be legally marketed in the U.S.

Food Contact Notifications

In 1997, the FD&C Act was amended through the passage of the Food and Drug Administration Modernization Act (FDAMA), which amended section 409 to allow for a premarket notification process as the primary mechanism for authorizing new uses of food contact substances (9). Prior to the establishment of the food contact notification (FCN) process, a food additive regulation was necessary to authorize the new use of new food contact substance. Importantly, there are two main similarities between the petition and notification processes and two notable differences. First, in both processes, the burden of proof for safety of the new intended use of an additive remains on the interested party (the petitioner in the case of a petition or the notifier in the case of a notification). Secondly, the same standard of safety applies to both submission types. However, unlike the petition process, the premarket notification process does not require the agency to publish an order in the Federal Register announcing the agency's safety decision and an authorizing regulation, if appropriate. Furthermore, unlike a food additive approval, the approval of a food contact substance through the notification program only applies to the manufacturer or supplier identified in the notification.

GRAS Substances

Under section 201(s) of the FD&C Act, a substance that is GRAS for a particular use in food is exempt from the definition of a food additive, and may lawfully be used without FDA review and approval. General recognition of safety must be based only on the views of qualified experts. The basis of such views may be either: 1) scientific procedures or 2) in the case of a substance used in food prior to January 1, 1958, through experience based on common use in food. In addition, general recognition of safety requires common knowledge about the substance throughout the scientific community knowledgeable about the safety of substances added to food. A determination that a particular use of a substance is GRAS through scientific procedures requires both technical evidence of safety (technical element) and a basis to conclude that this technical evidence of safety is generally known and accepted (common knowledge element). Interested persons may voluntarily submit information on their GRAS determination to FDA for review through the FDA's voluntary GRAS notification program.

FDA's Approach to Regulation of Nanotechnology Products

The agency recognizes the importance of communicating with its stakeholders when it comes to emerging technologies, such as nanotechnology, due to the potential impacts new technologies may have on the safety of the regulated product areas. Altered physicochemical properties that allow for product development opportunities warrant further consideration to determine the impact those properties have on the safety or other applicable attributes of the product. Therefore, the agency outlined its thinking on considerations related to nanotechnology in a draft guidance document entitled "Draft Guidance for Industry, Considering Whether an FDA-Regulated Product Involves the Application of Nanotechnology" that was released in June 2011 (*10*). Importantly, this draft guidance does not establish a formal definition for nanotechnology; rather, it establishes a framework and set of overarching principles intended to guide in the development of future, product-specific guidance documents. The agency discussed the overarching principles as "points to consider" when deciding whether a regulated product involves nanotechnology.

Even though nanotechnology may allow for altered properties compared to conventionally manufactured products, the agency's approach to assessing the safety of products that contain nanomaterials or otherwise involve the application of nanotechnology does not differ. The regulatory approach remains scientifically driven and risk-based where all products are required to meet the same legal standards applicable to each product type under the appropriate statutory authority. For example, in the case of food ingredients, the safety standard remains "reasonable certainty of no harm." While information relating to the application of nanotechnology to food ingredients may inform a safety decision, the agency does not a *priori* judge the use of nanomaterials or the application of nanotechnology in FDA-regulated products as inherently benign or harmful.

Materials at the nanoscale may exhibit altered properties compared to conventionally scaled counterparts. Typically, materials are manipulated to the nanoscale to bring about desirable changes offering new possibilities for innovation. For example, desirable changes to the substance such as an increase in bioavailability of a drug or an improvement to the taste, color, flavor, texture, or consistency of food may raise questions about the regulatory status, safety, effectiveness, or health impact of those materials. Size is a common characteristic that is used to describe a nanomaterial, nanoparticle, or nanotechnology with one nanometer (nm) being equal to a billionth of a meter. While definitions do exist for nanotechnology (e.g., the National Nanotechnology Initiative (NNI) Program defines nanotechnology as "the understanding and control of matter at dimensions between approximately 1 and 100 nm, where the unique phenomena enable novel applications") (*11*), FDA has not established a formal definition. Rather, the agency put forth two points to consider. The agency's "points to consider" are intended to be applicable to all FDA-regulated products and allow for the agency to maintain its existing product-focused, science-based regulatory policy frameworks. The agency recognized that size is one of the two aspects to consider when determining whether an FDA-regulated product contains nanomaterials or otherwise involves the application of nanotechnology; however the agency's interest extends beyond size alone and considers intentional manufacturing to achieve the desired properties. Furthermore, the agency recognizes the potential difference between products that have background levels of nanomaterials and those that have been deliberately manipulated or engineered to control particle size with the intent to generate size-related properties.

Ultimately, consideration of the size (i.e., one dimension in the nanoscale range, approximately 1–100 nm and up to 1 micron) and whether the engineered material exhibits properties or phenomena attributable to dimension allow FDA to take a broad, inclusive approach to considering whether FDA-regulated products contain nanomaterials or otherwise involve the application of nanotechnology.

More importantly, these criteria are general enough to be applicable to the existing product-specific statutory frameworks and are relevant to all FDA-regulated product areas. This adaptive approach offers flexibility to modify the presumptions or pathways for regulatory components as knowledge is gained about the potential role and importance of dimensions in the characteristics exhibited by engineered nanomaterials. The agency's draft nanotechnology guidance encourages industry to consult with the agency early in the product development process to address any questions about the safety and regulatory status of their products.

The 2011 draft guidance document provided the context for product-specific approaches consistent within the applicable statutory frameworks. For example, where premarket authority exists, attention to nanomaterials is incorporated into the established procedures; however, if the FD&C Act does not require premarket review of a particular product, consultation with the agency is encouraged to reduce the risk to human or animal health. Though the lack of premarket approval for some FDA-regulated products limits FDA's ability to review safety data before a product enters commerce, under U.S. law it remains the responsibility of the manufacturer to ensure the safety of their product.

Product-Specific Draft Guidance for Food Ingredients

Consistent with the agency's nanotechnology draft guidance, additional guidance documents have been published with product-specific guidance. In 2012, the agency released, "Draft Guidance for Industry: Assessing the Effects of Significant Manufacturing Process Changes, Including Emerging Technologies, on the Safety and Regulatory Status of Food Ingredients and Food Contact Substances, Including Food Ingredients that are Color Additives" (12). This draft guidance document addresses the impact that significant manufacturing changes, including the application of nanotechnology to food ingredients (13), have on product safety and regulatory status. This manufacturing guidance affirmed that the regulatory authority for food ingredients and color additives is sufficiently robust and flexible to address nanomaterials or products that involved the application of nanotechnology. The expectation exists that under certain circumstances, a significant alteration in the manufacturing process could impact the safety and/or regulatory status of a food ingredient resulting in the agency requiring a new authorization in order to clearly establish the conditions under which a food ingredient is safe and lawful.

Manufacturing changes are relevant to the agency's overall safety assessment because those changes can impact the identity, purity, or properties of the food ingredient, which in turn can impact the safety of the finished product. Therefore, changes in manufacturing must be considered, especially where they are deemed significant. For example, if the particle size distribution of a component of a food contact substance were altered to shift more fully to the nanometer scale there is the potential to affect the particle's ability to migrate from the food contact substance into the food itself. If the particle size of a direct food additive were altered, the food ingredient may have a different pharmacokinetic profile, biodistribution, and/or toxicological profile.

Any food ingredient manufacturing change has the potential to be significant and raise questions about the regulatory status because there is the possibility that the identity or safety of the food ingredient will be substantially altered. This remains true for manufacturing changes that involve nanotechnology. New technologies may introduce issues that warrant additional or different evaluation during the safety assessment of a food ingredient. In the case of nanotechnology, the regulatory status is questioned because: 1) most authorizations do not include size-dependent specifications; and 2) the safety of nanoscale versions of compounds cannot be demonstrated solely on evidence from conventionally scaled counterparts (14) due to novel properties and physical characteristics of nanoscale materials.

Traditionally, food ingredient authorizations do not include size-dependent specifications, such as particle size, size distribution, and morphology. However, updated chemistry guidance was issued by FDA for food contact substances, direct food additives, and color additives in December 2007, March 2009, and July 2009, respectively, that recommend including data related to particle size, size distribution, morphology, and other properties, as appropriate, if particle size is important for the additive to achieve the intended technical effect (15–17). Furthermore, FDA recommends that the intended technical effect and use sections of a petition or notification include a statement on the intended technical effect with data that demonstrates the specific size-dependent properties of the additive that affect the functionality (e.g., use of silver nanoparticles due to antimicrobial properties). Characterization of food ingredients that contain nanomaterials should include data related to any unique features in addition to the standard information, such as identity and purity (18).

There is evidence that particle size, surface area, aggregation/agglomeration, or shape may impact various toxicology endpoints and, thus have the ability to potentially alter the safety of the nano-engineered food ingredient (19, 20). In certain cases it may be warranted to examine the effects of those manufacturing changes on properties, including the effects on bioavailability of the food ingredient and its transport along the alimentary tract. Therefore, when contemplating a significant change in manufacturing process that involves nanotechnology additional testing may be required to demonstrate safety. For example, if the manufacturing change shifts the particle size distribution more fully into the nanoscale range, then the safety assessment should be based on the nanoscale version of the food ingredient. Likewise, if changes to the manufacturing process are undertaken so that the food ingredient exhibits altered properties, additional or different testing methods may be necessary to demonstrate the safety of the food ingredient.

FDA does not prescribe the tests that must be done to demonstrate safety; rather, the agency has consistently taken the position that many scientifically valid types of data may properly support a finding that the proposed use of a food ingredient is safe. In practice, FDA has applied exposure and toxicological criteria that are appropriate for the time and appropriate for assessing the safety of a particular food ingredient. Therefore, the interested party can choose an appropriate method and provide a justification as to why that method is appropriate, which removes the barrier of meeting established requirements regarding methodology and also allows for innovation in methods and process development.information, such as identity and purity.

Toxicology Considerations

Changes in the manufacturing process may impact the uptake, absorption, and bioavailability of the food ingredient, which in turn has the potential to impact the overall safety assessment. At this time, the state of the science does not support broad toxicology guidance for substances produced through nanotechnology. While the agency is continuing to monitor the state of the science and plans to update existing guidance or issue new guidance as appropriate, interested parties should contact the agency for specific case-by-case guidance to ensure that the product meets applicable legal requirements, including the standards for safety.

Impact of Manufacturing Changes on Regulatory Status

For existing regulations for food additives or color additives, it is important to understand that the identity and conditions of use of the food ingredient as described in the administrative record are relevant; therefore, a food ingredient may not be within the scope of a regulation if the identity, manufacturing process, or conditions of use are not consistent with how FDA evaluated the substance. Furthermore, the food ingredient must be of appropriate food grade, which includes consideration of potential impurities introduced into the food ingredient by the change in manufacturing process. When attempting to determine compliance with an existing regulation, FDA suggests consulting with the agency about the conclusions reached and making an appropriate regulatory submission to FDA, as appropriate.

In the case of an effective FCN, the agency is of the view that any manufacturing change intended to produce nanoscale particles would be considered a significant change, if such particles were not part of the original FCN. Section 409(h)(2)(C) of the FD&C Act states that a food contact substance approval does not apply to a similar or identical substance manufactured or prepared by a person other than the manufacturer identified in the notification. Therefore, significant manufacturing changes (e.g., a change intended to produce nanoscale particles) would be considered substantive and a new FCN would be required.

As discussed previously, GRAS substances are exempt from the definition of "food additive." GRAS substances require technical evidence of safety and a basis to conclude that this evidence of safety is generally known and accepted. Currently, FDA is unaware of generally available safety data sufficient to serve as the foundation for a GRAS determination for a food ingredient that involves the application of nanotechnology. Furthermore, at this time, FDA believes that a food ingredient manufactured using nanotechnology would likely not be GRAS or covered by an existing GRAS determination for a related food ingredient manufactured without using nanotechnology.

Conclusion

FDA published a draft guidance document in 2012 that discusses the impact of significant manufacturing changes on the safety and regulatory status of food ingredients. This guidance document describes the factors manufacturers should consider when determining the effect of a significant change in the manufacturing process, including the application of nanotechnology, for food ingredients already in the market. Significant changes in the manufacturing process have the potential to impact the identity, safety, and/or regulatory status of a product. Manufacturers are strongly encouraged to consult with the agency prior to bringing such a product to market.

References

1. Advancing Regulatory Science at FDA: A Strategic Plan. http://www.fda.gov/ScienceResearch/SpecialTopics/RegulatoryScience/ucm267719. htm (last accessed January 13, 2014).
2. FDA is responsible for regulating human and veterinary drugs, biological products, medical devices, food, cosmetics, tobacco products, and products that emit radiation. FDA is also responsible for advancing the public health by accelerating innovations to make medicines more effective and providing the public with accurate, science-based information on medicines and food to improve their health. FDA plays a significant role in addressing the Nation's counterterrorism capability and ensuring the security of the food supply (3).
3. The United States Government Manual. http://www.usgovernmentmanual. gov/Agency.aspx?EntityId=Ldrc/ujFJeo=&ParentEId=+klubNxgV0o= &EType=jY3M4CTKVHY= (last accessed January 13, 2014).
4. Hamburg, M. A. *Science* **2012**, *336*, 299–300.
5. About the Center for Food Safety and Applied Nutrition. http://www.fda.gov/AboutFDA/CentersOffices/OfficeofFoods/CFSAN/default.htm (last accessed January 13, 2014).
6. Section 201(s) defines the term "food additive" as any substance the intended use of which results or may reasonably be expected to result in its becoming a component or otherwise affecting the characteristics of any food, if such substance is not GRAS. A substance is GRAS if it is generally recognized, among experts qualified by scientific training and experience to evaluate its safety, as having been adequately shown through scientific procedures (or, in the case of a substance used in food prior to January 1, 1958, through either scientific procedures or experience based on common use in food) to be safe under the conditions of its intended use.
7. Certain other substances that may become components of food are also excluded from the statutory definition of food additive, including pesticide chemicals and their residues, new animal drugs, color additives, and dietary ingredients in dietary supplements (21 U.S.C. 321(s)(1) through (s)(6)).
8. Section 201(t) defines the term "color additive" as a material which is a dye, pigment, or other substance made by a process of synthesis or similar artifice ... when added or applied to a food, drug, or cosmetic, or to the human body or

any part thereof, is capable (alone or through reaction with other substance) of imparting color thereto) except that such term does not include any material that is used solely for a purpose other than coloring.

9. Section 409(h)(6) defines the term "food contact substance" as any substance intended for use as a component of materials used in manufacturing, packing, packaging, transporting, or holding food if such use is not intended to have any technical effect in such food.

10. Draft Guidance: Considering Whether an FDA-Regulated Product Involves the Application of Nanotechnology. http://www.fda.gov/ScienceResearch/SpecialTopics/Nanotechnology/ucm257926.htm (last accessed January 13, 2014).

11. Nano.gov: National Nanotechnology Initiative. http://www.nano.gov/nanotech-101/what (last accessed January 13, 2014).

12. Draft Guidance for Industry: Assessing the Effects of Significant Manufacturing Process Changes, Including Emerging Technologies, on the Safety and Regulatory Status of Food Ingredients and Food Contact Substances, Including Food Ingredients that are Color Additives. http://www.fda.gov/Food/GuidanceRegulation/GuidanceDocumentsRegulatory Information/IngredientsAdditivesGRASPackaging/ucm300661.htm (last accessed January 13, 2014).

13. The term "food ingredient" in this document includes substances that are subject of a food additive or color additive regulation, food contact substance notification, or a GRAS determination.

14. Gonzales, L.; Lison, D.; Kisch-Volders, M. *Nanotoxicology* **2008**, *2*, 252–273.

15. Guidance for Industry: Preparation of Premarket Submissions for Food Contact Substances: Chemistry Recommendations. http://www.fda.gov/Food/GuidanceRegulation/GuidanceDocumentsRegulatoryInformation/IngredientsAdditivesGRASPackaging/ucm081818.htm (last accessed January 13, 2014).

16. Guidance for Industry: Recommendations for Submission of Chemical and Technological Data for Direct Food Additive Petitions. http://www.fda.gov/Food/GuidanceRegulation/GuidanceDocumentsRegulatoryInformation/IngredientsAdditivesGRASPackaging/ucm124917.htm (last accessed January 13, 2014).

17. Guidance for Industry: Color Additive Petitions – FDA Recommendations for Submission of Chemical and Technological Data on Color Additives for Food, Drugs, Cosmetics, or Medical Devices. http://www.fda.gov/ForIndustry/ColorAdditives/GuidanceComplianceRegulatoryInformation/ucm171631.htm (last accessed January 13, 2014).

18. The identity of a food ingredient is usually described in terms of information such as: name (chemical name or common trade name); applicable identification number (e.g., Chemical Abstracts Service Registry Number); applicable chemical formula; source (if of natural biological origin); quantitative composition; impurities and contaminants; and physical and chemical properties (e.g., melting point, boiling point, specific gravity,

refractive index optical rotation, pH, solubility, reactivity, particle size, and chromatographic, spectroscopic or spectrometric data).

19. European Food Safety Authority. *The EFSA Journal* **2009**, *958*, 1–30.
20. Scientific Committee on Emerging and Newly Identified Health Risks. http://ec.europa.eu/health/scientific_committees/opinions_layman/nanomaterials/documents/nanomaterials.pdf (last accessed January 13, 2014).

Chapter 5

A Comprehensive Study into the Migration Potential of Nano Silver Particles from Food Contact Polyolefins

J. Bott,* A. Störmer, and R. Franz

Fraunhofer Institute for Process Engineering and Packaging IVV,
Giggenhauser Straße 35, 85354 Freising, Germany
*E-mail: johannes.bott@ivv.fraunhofer.de

The potential of nano silver particles (Ag-NPs) to migrate from food contact polyolefins into food was systematically investigated. Migration studies were carried out using low density polyethylene (LDPE) films with different concentrations of incorporated Ag-NPs in contact with different EU-official food simulants simulating long-term storage with aqueous and fatty food contact. Detectable migration of total silver as measured by inductively coupled plasma mass spectrometry (ICP-MS) was found in aqueous food simulants only. Stability tests of Ag-NPs in these food simulants by asymmetric flow field-flow fractionation (AF4) analysis showed rapid oxidative dissolution of the Ag-NPs and demonstrated that only ionic silver was present in the migration solution. Non-detectability of silver both in the isooctane and 95 % ethanol migrates indicated that Ag-NPs would not be able to migrate. These findings were supported by a new approach of migration modeling showing that nanomaterials (NMs) in general are immobilized in a polymeric matrix, resulting in a very limited hypothetical potential for the migration of NMs smaller than 3-4 nanometer in diameter. However, such small nanoparticles are usually not found in polymer nanocomposites. The results of this study suggest that migration of nanoparticles from food contact plastics cannot lead to an exposure of the consumer.

Introduction

Silver nanoparticles (Ag-NPs) are fine particles of elemental silver. Commonly they are produced by the reduction of positively charged silver ions (Ag^+) in an aqueous solution, resulting in elemental silver (Ag^0). In a further step of production the silver atoms aggregate to form silver particles, whereby the growth of particles can be controlled by the addition of stabilizers, resulting in spherical Ag-NPs with consistent particles sizes (1–3). Ag-NPs are primarily used as an antimicrobial agent. Silver is already known for a long time for its effectiveness against microbes. The mode of action for the antimicrobial effect is caused rather by the release of Ag^+ ions than by Ag^0 itself (4). In comparison to silver as a solid bulk material, Ag-NPs promise even higher antimicrobial activity because of the increased surface to volume ratio (1) that is typical for all kinds of nanomaterials (NMs) and which provides a higher potential to release Ag^+ ions. Many publications address the toxicity of Ag-NPs (5–7) and the mechanism of the antimicrobial activity supporting that the effect is caused by the release of Ag^+ from the surface of the Ag-NPs (8–11). Because of the higher efficiency, Ag-NPs are used in a variety of products for medical applications (12) and also for many consumer products like refrigerators, textiles and cosmetics (1, 4, 13). A promising application area of Ag-NPs seems to be the food packaging sector where it has a potential to be used in polymer nanocomposites for food contact materials (FCMs) (14–16). Nanocomposites using Ag-NPs belong to the so-called active packagings, whereby the packaging is able to interact with the environment of the packed food. In case of Ag-NPs the benefit is seen in enhanced shelf life of the packed food, based on the antimicrobial activity of the released Ag^+.

On the other side, apart from the advantages, public concern about NMs in consumer products in general and in food packaging in particular can be perceived throughout the last 5 years (17, 18). There is still a lack of knowledge regarding the exposure and uptake of NMs by the consumer which drives the authorities to make use of the precautionary principle and handle the issue very conserv-atively. In case of plastic FCMs the European Union (EU) requires in the EU Plastics Regulation 10/2011 (19) that NMs must be specifically approved when they are used in their 'nano'-form. Ag-NPs are not included in this regulation so far. Regarding safety evaluation of Ag-NPs in FCMs a crucial point is the knowledge about the migration potential of this substance. Although Ag-NPs belong to the most frequently investigated NMs, only few studies address the migration of either total Ag or nano Ag out of polymeric packaging materials. Cushen et al. (20) report on the potential migration of Ag-NPs from a PVC nanocomposite and Song et al. (21) on total Ag migration from a nanosilver-polyethylene composite. Goetz et al. (22) and Huang et al. (23) claim to have measured migration of Ag-NPs from commercially available food containers. Whilst the first two studies measured and report on total silver migrating out of the polymer, the latter two studies report migration of nanoparticulate Ag into the used food simulants, after contact with the food container. The two latter findings are not in line with modern scientific knowledge about migration and contradict the expectation from a physico-chemical point of view. Migration of additives homogeneously distributed in a polymer follows in general Fick's law of diffusion where the diffusion within the polymer matrix strongly depends on the size of the migrant. Ag-NPs, typically used at sphere diameters of at least 20 nm are by far larger than usual conventional plastics additives such as antioxidants. From this point of view, it appears that Ag-NPs seem to be too large to exhibit a noticeable mobility within a polymeric matrix. Such expectations were also expressed in a paper by Simon et al. (*24*).

The objective of our study was therefore to investigate systematically and comprehensively into this issue. The idea was to carry out conclusive migration tests, which means in a time dependent mode using different food simulants at different temperatures and on LDPE samples containing different concentrations of Ag-NPs. Furthermore, from an analytical point of view this means applying besides element-specific analysis, like inductively coupled plasma mass spectrometry (ICP-MS), also asymmetric flow field-flow fractionation (AF4) analysis as a direct measurement method. With the above expressed expectation we were aware of the difficulties to measure Ag-NPs directly at sufficiently low concentrations or detection limits. Therefore, we established also a migration model to provide more clarity in this difficult question. Furthermore, our objective was to study also the time dependent oxidative dissolution of Ag^0 in contact with aqueous and acidic food simulants to substantiate whether or not Ag-NPs migration would be measurable at all in these media.

Materials and Methods

Materials

Low density polyethylene (LDPE) films with three different concentrations of silver in the polymer (LDPE A, B and C) were extruded using a Collin flat film extruder (Dr. Collin GmbH, Germany). A nanosilver containing masterbatch ROMBEST AM 6500NANO (Romcolor, Romania), which contained 6500 mg kg^{-1} silver was mixed with a B21/2.0 LDPE (Rompetrol Petrochemicals, Romania) and was extruded to films of 60 µm thickness. Additional blanks without silver were produced in the same way as reference (LDPE 0). Thus, LDPE films with a nominal concentration of 0 mg kg^{-1} (LDPE 0), 50 mg kg^{-1} (LDPE A), 150 mg kg^{-1} (LDPE B) and 250 mg kg^{-1} (LDPE C) silver were produced. PL-Ag-S10 colloidal silver dispersion (Plasmachem GmbH, Germany) was used for stability tests of Ag-NPs. This colloidal silver dispersion had a concentration of 100 mg l^{-1} silver, with particles of about 10 nm in diameter.

Transmission Electron Microscopy (TEM)

TEM images of the polymeric films with the lowest and highest concentration of Ag-NPs (sample A and C) were prepared by Innoform GmbH, Germany. With this technique the distribution and size characteristics of the Ag-NPs in the polymer can be visualized. For sample preparation the polymeric films were subjected to cryo-ultra-thin-sectioning using a diamond knife.

Inductively Coupled Plasma Mass Spectrometry (ICP-MS)

ICP-MS measurements were carried out using a 7700 series ICP-MS (Agilent Technologies, USA) to determine the amount of silver in the migration samples and the silver content of the LDPE films. The setup of the ICP-MS is summarized in Table I. The ICP-MS was calibrated using MERCK VI multi-element ICP-MS calibration standard (Merck KGaA, Germany), which was diluted with 3 % nitric acid to silver concentrations of 1, 2, 5, 10, 50, 100 and 200 µg l^{-1}.

Table I. Setup of the ICP-MS.

RF Power	1550 W
Plasma Gas	Argon, 15 l min^{-1}
Peristaltic pump speed	0,3 rps
Nebulizer	Micro Mist (Agilent)
Autosampler	ASX-520 (Agilent)
Measuring Mode	Helium and No Gas
Measured isotope	Ag^{107}

Determination of the Silver Content of the LDPE Films by Acid Digestion

The LDPE films with nominal 50, 150 and 250 mg kg^{-1} Ag-NPs and the LDPE blank were digested to determine the content of silver. Each sample film was cut into small pieces of about 3x3 mm. About 20 mg were weighed out into polytetrafluoroethylene (PTFE) cells and filled with 10 ml of 69 % nitric acid (J. T. Baker, for trace metal analysis). The cells were then stored in pressure vessels for 10 h at 160 °C in a temperature controlled oven. At the end of storage 0.5 ml of each sample was diluted with 9.5 ml ultra-pure water and analysed for silver by ICP-MS. For the determination four identical samples of each material were prepared.

Migration Experiments

The films LDPE A, B, C and 0 were stored in 3 % acetic acid, 10 % ethanol and 95 % ethanol by total immersion at 60 °C for 3, 6, 8 and 10 days. Additionally isooctane was chosen as second alternative fat simulant under rapid ex-traction conditions at 40 °C for 24 hours according to EN 1186-15. The migration experiments were performed according to EN 13130-1 and EN 1186-3. An area of 1 dm² of the respective LDPE film was stored in 100 ml Schott-bottles with PTFE-sealed closure. The bottles were filled with 100 ml of the respective simulant. At this volume the test films were completely covered with the simulant. The samples were then stored in a temperature controlled oven. At the end of the storage time the films were removed from the simulant and the simulant was transferred in several steps quantitatively to 15 ml centrifuge tubes. All simulants were evaporated carefully under a gentle nitrogen stream (at 40 °C) to dryness. Then the residues were taken up in 10 ml 3 % nitric acid for ICP-MS measurements.

For each simulant recovery check samples were prepared equally. Simulants were fortified to 100 µg l⁻¹ silver by adding 0.1 ml of the 100 mg l⁻¹ colloidal silver dispersion in a 100 ml volumetric flask and filling it up with the respective simulant. The recovery check samples were then transferred into Schott-bottles and stored under the same conditions as the migration samples, with the same subsequent sample work-up procedure. The detected amount of silver was compared to a freshly prepared 100 µg l⁻¹ standard in 3 % nitric acid. Both, the migration samples and the recovery check samples were done in triplicate.

Asymmetric Flow Field-Flow Fractionation (AF4)

AF4 measurements were carried out with 'AF2000 MT Series mid temperature' (Postnova Analytics, Germany). The system was equipped with a 350 µm channel and a regenerated cellulose membrane (cut-off: 10 kDa, Postnova Analytics). The channel was constantly tempered to 40 °C. Water from the ultra-pure water system TKA GenPure (TKA, Germany) stabilized with 100 mg l⁻¹ sodium azide was used as flowing liquid for the AF4. The channel flows were controlled by the software 'AF2000 Control Program' (Postnova), using the following flow program: Samples were injected into the channel with an injection flow of 0.2 ml min⁻¹ and focused with a focus flow of 2.15 ml min⁻¹. During the injection time of 10 min and an additional transition time of 0.2 min, the cross flow was kept constant at 1.85 ml min⁻¹. After transition (i.e. deletion of the focus flow) the cross flow was reduced within 0.2 min to 1.0 ml min⁻¹ followed by a non-linear decline to 0.08 ml min⁻¹ within 20 min using a power gradient of 0.2. The cross flow was then kept constant for additional 10 min. At the end the channel was flushed for 15 min without any cross flow. The main channel flow was kept constant at 0.5 ml min⁻¹ for the whole run. Samples were injected by full loop injections of a PN5300 series autosampler (Postnova Analytics) equipped with a 1000 µl sample loop.

For stability tests, colloidal silver dispersions were prepared in 3 % acetic acid. First the 100 mg l⁻¹ dispersion was diluted to a silver concentration of 1.0 mg l⁻¹ with ultra-pure water. The diluted colloidal silver dispersion was then used to prepare a 50 µg l⁻¹ silver dispersion by adding 1.0 ml into a 20 ml volumetric flask and filling it up with 3 % acetic acid. The Ag-NPs dispersion was injected into the AF4 directly after preparation and was then stored at room temperature. AF4 measurements were repeated each hour for a time span of 5 hours. For stability tests of colloidal silver in ultrapure water, the 1 mg l⁻¹ colloidal silver dispersion in ultrapure water was measured directly after preparation and again after 24 hours stored at room temperature.

Multi Angle Laser Light Scattering Spectrometry (MALLS)

A 21-angle MALLS detector PN3621 (Postnova Analytics, Germany) was used to record the signal curve of Ag-NPs injected into the AF4 system. The device was controlled by the software 'AF2000 Control Program' (Postnova). The detector was connected directly behind the AF4 system and was operated at $\lambda = 532$ nm at 12.5 mW laser power. Except for the lowest detector angle ($7°$) the outputs of all other MALLS angles were integrated with an Excel-tool and summed up to obtain the total peak area. This was used to correlate the MALLS outputs to the amount of Ag-NPs injected into the AF4. Particle size calculations were performed by the software 'AF2000 Control Program' (Postnova Analytics, Germany).

Migration Modelling

The applied migration model for estimation of migration of nano particles from polyolefins is based on Fick's 2nd equation (equation 1).

$$\text{Eq. 1:} \qquad \frac{\partial c_P}{\partial t} = D_P \frac{\partial^2 c_P}{\partial x^2}$$

where: c_P is the concentration of the migrant in the plastic at time t at distance x from the interface between food and plastic.

This differential equation can be analytically solved for migration from a monolayer material into a well mixed liquid (*25, 26*) resulting in a function for the surface area related migration depending on the diffusion coefficient D_P, the partion coefficient $K_{P,F}$ between polymer and food, the thickness and density of the polymer as well as the packaging geometry (volume food and volume polymer). Todays software solutions use numerical mathematics to solve the differential equation (*27, 28*).

The kinetic part is represented by the temperature dependent diffusion coefficient D_p. The diffusion coefficient $D_{p,i}$ of a substance i in the polymer P was calculated by equation 2 developed by Piringer (*29, 30*).

$$\text{Eq. 2:} \qquad D_{P,i} = D_u \exp(w_{i,e} - w_{p,e} \bullet 0.14(14j+2)^{2/3} - ww_{j,e}^{2/3} T_{m,p} R/RT)$$

where: $w_{i,e} = (1+2\pi/i)^{i/e}$, $j = (i^{1/3})$, $w_{j,e} = (1+2\pi/j)^{j/e}$, $p = (M_{r,p}/14)^{1/3}$, $w_{p,e} = (1+2\pi/p)^{p/e}$, $i = (M_{r,i} - 2)/14$, $w = e^{2\pi}/e$ and $D_u = 1$ m²s^{-1}

This equation 2 is derived from equations describing diffusion in liquids but considers the interaction of the migrating substances with the polymer matrix by the interaction parameter $w_{p,e}$. This parameter includes the relative molecular mass of the polymer, $M_{r,p}$, and its melting temperature, $T_{m,p}$. The parameters $w_{i,e}$ and j are functions of the molecular weight of the migrating substance $M_{r,i}$. R is the gas constant and T the absolute temperature in Kelvin. More detailed explanations can be found elsewhere (*29*).

In our case of nanoparticles as potential migrants, the mass of the particle of interest was calculated and considered as quasi-molecular weight $M_{r,i}$ of this particulate migrant. For LDPE as a worst case polyolefin (highest diffusion characteristics) the polymer interaction parameter was calculated with $T_{m,p} = 110$ °C and $M_{r,p} = 5000$ g mol^{-1}. This $M_{r,p}$ value represents a further conservatism because LDPE has usually higher molecular weight averages which would lead to even lower diffusion coefficients for migrants.

The calculation of the migration was performed using the MIGRATEST®EXP version 2011 software (*31*) and for diffusion coefficients lower than 10^{-25} cm² s^{-1} using AKTS software (*32*). Both softwares use numerical algorithms for migration calculation.

Results

Characterization of the Nanomaterial in the Polymer by TEM

TEM images were taken of test samples LDPE A and LDPE C. At lower magnification (Figure 1) quantitative size characterization of the incorporated Ag-NPs could not be visualized. However, it appears that a homogenous distribution was achieved in the polymer which can be expected because of the homogenisation before and during the extrusion process. At higher magnification the Ag-NPs could be clearly visualized as particles at nanoscale dimensions (Figure 2). In both polymer films (LDPE A and LDPE C), Ag-NPs with different particle sizes could be found. The smallest particles found were shaped spherical with diameters of about 10 nm. The largest particles or aggregates found showed a more ellipsoidal shape with diameters from 100 to 270 nm. Also some rod-like particles with dimensions about 35 x 200 nm were detected. Most commonly, however, spherical particles with about 50 nm in diameter were found in both films. It is justified to assume that test sample LDPE B has the same Ag-NPs characteristics.

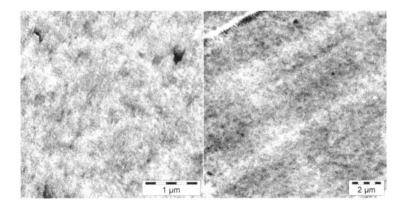

Figure 1. TEM images of the 50 mg kg^{-1} silver in LDPE film (left) and the 250 mg kg^{-1} silver in LDPE film (right). Images taken at low magnification.

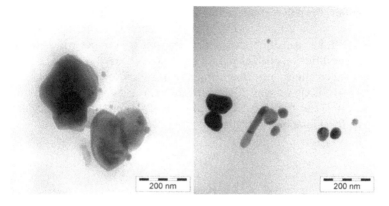

Figure 2. TEM images of the 50 mg kg⁻¹ silver in LDPE film (left) and the 250 mg kg⁻¹ silver in LDPE film (right). Images taken at high magnification.

Determination of the Silver Content of the LDPE Films by Acid Digestion

The total amount of silver in the four LDPE films was determined by acid digestion and subsequent ICP-MS measurements (Table II). The variations of the silver concentration in the three test specimens of LDPE films A, B and C, expressed by the relative standard deviations, indicate that the Ag-NPs are homogeneously distributed in the polymer matrix. LDPE film B showed a somewhat higher variation of the silver content of the three investigated specimens.

Table II. Silver concentration of the different LDPE films

Sample	Nominal concentration of silver [mg kg⁻¹]	Silver concentration measured by ICP-MS [mg kg⁻¹]
LDPE 0	0	0.1 ± 0.03
LDPE A	50	48.7 ± 2.6
LDPE B	150	185.2 ± 27.4
LDPE C	250	249.8 ± 5.7

Recovery Experiments and Detection Limit of Silver by ICP-MS

The comparison of the recovery check samples with the reference sample showed that all used food simulants are suitable for the sensitive detection of silver. For all food simulants between 85 % and 95 % of the added silver could be recovered after storage for 10 days at 60 °C which is a very satisfying recovery criterion. The recovery rates which were 84.7 % for isooctane, 85.9 % for 95 % ethanol, 91.7 % for 10 % ethanol and 94.4 % for 3 % acetic acid, were taken into account to calculate the overall method detection limit. Under consideration of the used sample volume and recovery rates the detection limits for the migration experiments expressed as mass Ag per area of LDPE film were 11.8 ng dm^{-2} for isooctane, 11.6 ng dm^{-2} for 95 % ethanol, 10.9 ng dm^{-2} for 10 % ethanol and 10.6 ng dm^{-2} for acetic acid. The recovery check samples were prepared by fortifying the respective simulant with a colloidal Ag-NPs standard, without using any additive for the stabilization of the Ag-NPs. Dissolving of Ag-NPs in aqueous simulants, like 3 % acetic acid and 10 % ethanol, led to higher recovery rates, whereas in organic simulants, like 95 % ethanol and isooctane, the attachment of particles on the glass surface might be the reason for loss of sample.

Migration of Silver from the LDPE Films

In the aqueous food simulants 10 % ethanol and 3 % acetic acid silver migration was clearly measurable and found to be dependent on the initial silver concentration in the polymer (Figure 3). Migration of total silver after 10 days at 60 °C was 2.4 µg dm^{-2} (LDPE A), 13.2 µg dm^{-2} (LDPE B) and 115.1 µg dm^{-2} (LDPE C) into 10 % ethanol and 168.5 µg dm^{-2} (LDPE A), 444.8 µg dm^{-2} (LDPE B) and 1010.9 µg dm^{-2} (LDPE C) into 3 % acetic acid. In migration control samples prepared with LDPE blanks (LDPE 0) silver was not found in any of the respective simulants. ICP-MS measurements of migration samples from all three test films obtained from fatty food simulants isooctane and 95 % ethanol did not show detectable silver amounts thus indicating that no silver was released from the polymer matrix even after 10 days at 60°C in 95 % ethanol and 24 hours at 40°C for isooctane.

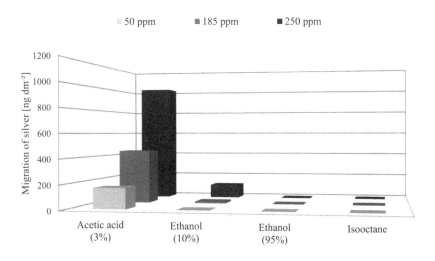

Figure 3. Results of the migration measurements. Samples stored for 10 days at 60 °C (3 % acetic acid, 10 % ethanol, 95 % ethanol) and 24 hours at 40 °C (isooctane).

Besides a concentration (in the LDPE film) dependent release of silver into the aqueous food simulants a time dependent migration of silver could be observed by the kinetic setup of the experiment (Figure 4 and Figure 5). In both cases migration appears to be at maximum from LDPE films A and B after 10 days and to further increase from LDPE C film.

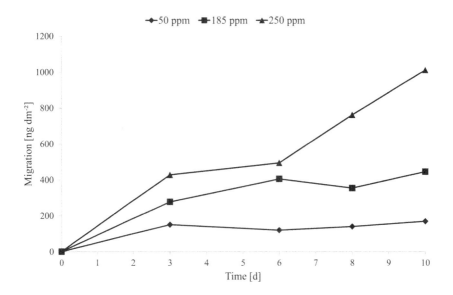

Figure 4. Migration of silver into 3 % acetic acid, stored at 60 °C.

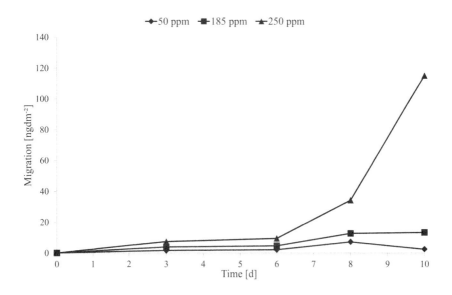

Figure 5. Migration of silver into 10 % ethanol, stored at 60 °C.

Stability Testing of an Ag-NPs Dispersion

An AF4/MALLS run of a freshly prepared 50 μg l^{-1} Ag-NPs dispersion caused a clear signal in the AF4 fractogram. The Ag-NPs of the freshly prepared dispersion eluted from t = 13 min until the separation force was zero at t = 40 min. Under assumption of a compact spherical structure of these Ag-NPs, the particle diameter was calculated to be 10.8 nm, using a Zimm plot. The same sample was injected again hourly for total time period of five hours. Already after one hour at room temperature a significant decrease of the MALLS signal could be recorded. With each injection a continuous decrease of the signal was observed until after 5 h the signal of the Ag-NPs dispersion could not be differentiated from the 3 % acetic acid blank anymore (Figure 6).

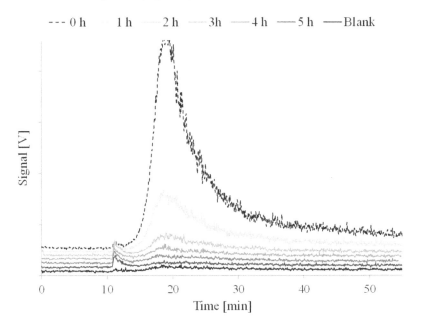

Figure 6. Signals of the 92° MALLS detector of a 50 μg l^{-1} Ag-NPs standard stored in 3 % acetic acid. The signal curves were shifted by 40 mV to each other for better visibility.

For a better correlation of the signal intensities to the injected amount of Ag-NPs, the total peak area was calculated. Therefore, the signal intensities of all MALLS angles (with exception of the 7 °angle) were integrated and summed up using an Excel tool. By plotting the total peak area of each sample in a logarithmic scale versus the storage time (Figure 7) the kinetic degradation of Ag-NPs becomes visible and can be correlated to the time. From the slope a half-life of the Ag-NPs in 3 % acetic acid of 0.6 h can be derived. In contrast to that, Ag-NPs stored in ultrapure water showed a higher stability compared to 3 % acetic acid. 1 mg/l colloidal silver dispersed in ultrapure water was stored under the same conditions.

After 24 h at room temperature more than 80 % of the silver particles still could be recovered, i.e. about 20 % of the silver was oxidized to Ag^+.

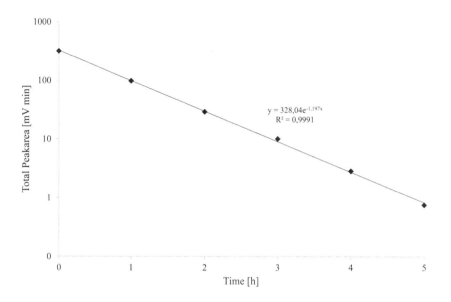

Figure 7. Decrease of the total peak area (measured by AF4/MALLS) by resolving of Ag-NPs in 3 % acetic acid.

Migration Modelling

It is generally recognised that the diffusion coefficient of a migrant in a polymer is mainly a function of the molecular volume of the diffusing species. This has in particular been shown in a recent publication on the diffusion of chemical substances in polyethylene terephthalate polymer (*33*). In this paper, correlations between calculated molecular volumes of migrants and their experimentally determined diffusion coefficients as well as activation energies of diffusion are presented. Due to the better accessibility of the molecular weight of a known chemical, generally recognised diffusion models (*28, 34, 35*) which are used for conservative migration modelling the molecular weight is used as the main criterion and estimator for the molecular size of the migrant.

Small NPs are not really molecules with a defined molecular weight. On the other hand when considering NPs as potential migrants, their potential to move across the polymer network will also be determined by the size or better the free cross section of the particle. In order to construct a relationship between the size of a NP and the molecular weight which would then allow using equation 2 of the migration model, the NP is considered as a quasi-molecule with the consequence that the mass of the NP is taken as its molecular weight.

Table III. Overview of physico-chemical specifications calculated for spherical carbon nanoparticles (assumed density 1g cm^{-3}) of diameters d = 1 nm to d = 10 nm.

Diameter d [nm]	Volume [nm^3]	Mass [E-21 g]	Number C atoms	Quasi Mol. weight [g mol^{-1}]	Diff. coeff. D_{LDPE} [cm^2 s^{-1}][a]
1	0.52	0.52	26	314	4.3 E-9
1.5	1.77	1.77	88	1060	1.0 E-11
2	4.19	4.19	209	2512	8.3 E-14
2.5	8.18	8.18	409	4906	1.2 E-15
3	14.13	14.13	707	8478	2.3 E-17
3.5	22.44	22.44	1122	13463	5.5 E-19
4	33.49	33.49	1675	20096	1.6 E-20
4.5	47.69	47.69	2384	28613	6.2 E-22
5	65.42	65.42	3271	39250	2.1 E-23
6	113.04	113.04	5652	67824	4.2 E-26
7	179.50	179.50	8975	107702	1.2 E-28
8	267.95	267.95	13397	160768	1.8 E-31
9	381.51	381.51	19076	228906	2.0 E-33
10	523.33	523.33	26167	314000	1.1 E-35

[a] At 40 °C.

In this study modeling was performed using carbon black as example. If we assume that the NPs are spheres (highest mobility in polymer) and consist chemically of carbon as an element with a low atom weight (12 g mol^{-1}) and would have a low density of 1 g cm^{-3} then for a given size (diameter d) of the sphere the calculated quasi molecular weight of this NP would represent the worst case migrant compared to any other NP migrant of the same size but consisting of a different chemical composition. With other words, the diffusion coefficients calculated from these molecular weights for LDPE from equation 2 would overestimate the diffusion and migration of any other NPs of the same size in polyolefins.

Based on these assumptions, for spherical NPs with diameters from d = 1 nm up to d =10 nm the corresponding volumes and masses as well as numbers of C atoms and quasi-molecular weights have been calculated. The compilation of these data is listed in Table III. In the last column of Table III, diffusion coefficients at T= 40 °C have been calculated from equation 2. It can be seen that NPs with d = 1 nm (m.w. = 314 g mol^{-1}) and d = 1.5 nm (m.w. = 1060 g mol^{-1}) would fall into the molecular weight category of usual plastics additives. From d = 2 nm onwards the NP mass increases and thus the diffusion coefficients decrease rapidly in an exponential way. At d = 10 nm the NP would move in LDPE with a diffusion coefficient of 1.1 E-35 cm² s⁻¹ which is an extraordinary low value and never observed from any plastics additive in any polymer matrix.

Using equation 2, the diffusion coefficients from Table III can be used to model migration of the NPs from LDPE. The following scenario is assumed: a NP of a given diameter should be present in LDPE at a concentration ($C_{P,0}$) of 2.5 % (25000 mg kg⁻¹). This $C_{P,0}$ was taken from practical reasons because the maximum use level of carbon black additive in polymers is 2.5 % according to the European Legislation (19). Migration for each size category of NP from an LDPE layer (assumption for thickness: 3 mm) into a food simulant (FS) which disperses the NPs very well (assumption for $K_{LDPE/FS}$ = 1) after 10 days at 40 °C can be then modeled as depicted in Figure 8.

For better understanding of the Figure 8: if the 2.5 % NM would consist of 1 nm NPs only then the migration value would be about 770 mg kg⁻¹, if it would consist of 10 nm NPs only then the migration would be 3.5 E-17 mg kg⁻¹. Of course, as can be seen from the TEM measurements there is a NP size distribution where these small NPs from 1 to 10 nm were not observed. When the size distribution of NPs in a polymer is known then according to the size dependent $C_{P,0}$ values a differentiating migration modelling could be carried out. As already indicated above NPs with d = 1 nm and d = 1.5 nm for which D values as for usual plastics additives can be derived, these NPs would consequentially give migration values as usual additives. For NP sizes of d = 3.5 nm and d = 4.5 nm migration values in the µg kg⁻¹ respectively ng kg⁻¹ range can be modelled. For sizes from d = 5 nm onwards migration values are decreasing exponentially down to meaningless concentrations.

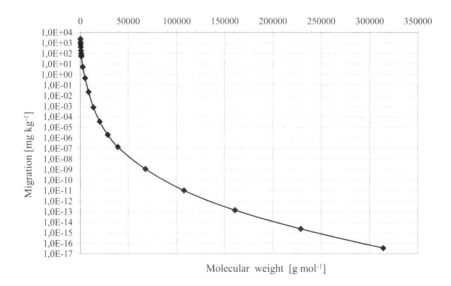

Figure 8. Modeled migration (10 days at 40 °C) of spherical carbon NPs from LDPE (d= 3 mm, $C_{P,0}$ = 2.5 %) as a function of their molecular weight respectively sphere diameter.

Discussion

The key objective of this study was to examine if silver can migrate in its elemental nano form out of a polyolefine matrix. For this a clear analytical differentiation between silver ions and Ag-NPs is essential. The antimicrobial activity from Ag-NPs containing polymeric packaging materials is based on the release of silver ions which are formed by oxidative dissolution of Ag^o and for which suitable contact media are required (*11*). Kittler (*36*) demonstrated in a comprehensive work that the silver ions diffuse faster than they are detached from the surface of the Ag-NPs. He could demonstrate that this was strongly depending on the used contact medium. Therefore, the dissolution of Ag-NPs may be the rate-determining parameter for the migration of silver out of nanocomposites. The results of our study show that silver migrated out of the LDPE matrix when the polymer was in contact with aqueous food simulants. However, when organic simulants were used, no release of silver from the polymer could be found, although our method validation showed that recovery was very satisfying and detection of silver in these types of food simulants was possible at very low detection limits. In our work silver was only found in those food simulants in which Ag-NPs are chemically unstable and in which ionic Ag species can be dissolved. Migrated silver was detected in traces in 10 % ethanol. This can be related to the high water content of this simulant leading to dissolution of ionic Ag from the Ag-NPs. As shown by stability experiments of Ag-NPs in ultrapure water ionic silver was formed, but in a lower extend than in the acidic medium 3 % acetic acid. 3 % Acetic acid caused a significant migration of silver about nine times higher than into 10% ethanol. This is chemically plausible when considering that 3 % acetic acid is known to be the strongest food simulant for dissolving inorganic species and to oxidatively dissolve Ag-NPs. In contrast to that, the organic simulants 95 % ethanol and isooctane did not show any release of silver. Isooctane is known to produce stable Ag-NPs dispersions (*37*) and ethanol is able to reduce silver ions (*38*) for what reason both simulants rather stabilize Ag-NPs than dissolve them. As a consequence, migration values obtained in the two aqueous simulants are the most relevant ones with regard to the key objective of our study.

The chemistry related assumption, that Ag-NPs are not stable in 3 % acetic acid due to oxidative dissolution could be demonstrated clearly by AF4/MALLS measurements. A fundamental prerequisite for AF4/MALLS measurements is that the sample contains stable particles (*39, 40*). If particles are dissolved, they will be washed out (*41*). In general AF4 is a suitable technique for the separation of particles in the size range of 1 nm to several microns (*42*), but silver ions with an effective ion radius of 0.115 nm (*43*) are too small. For our AF4/MALLS measurements we used Ag-NPs with 10 nm in diameter since those particles were the smallest found in the LDPE films. They can be considered as those with the highest migration potential out of all present in the polymer. Injections of Ag-NPs dispersions which had been stored for five hours at room temperature in 3 % acetic acid did not cause a signal anymore. The stability test using ICP-MS had shown that the total silver concentration remained constant even after 10 days at 60 °C. This makes obvious that the Ag-NPs were completely dissolved within this short time span. It can be reasonably expected that particles when stored longer and even at higher temperatures will be dissolved too, even larger ones. Therefore and from the considerations made above, it becomes quite obvious that the detected silver in 3 % acetic acid was in ionic form. The same is the case for silver in 10 % ethanol. With other words and more stringent: it appears to be impossible to measure migration of silver nanoparticles as an end parameter from nano-silver containing polyolefins.

We are aware that we may be in conflict with results reported on this issue in other published studies. We found two studies that report explicitly nano silver migrating out of the polyolefin samples. However, both studies have in common that they used different approaches for the differentiation of silver ions and Ag-NPs, and both did not differentiate the silver species directly in the migration samples where the silver was detected. The system Ag^0/Ag^+ is very sensitive to oxidation and reduction, thus it strongly depends on the reaction partners during sample preparation which species will be found. Huang et al. (*23*) performed migration experiments on commercially available food containers and reported migration of nano silver. The results of this study are in so far surprising that the migration of silver was independent from the used food simulant. In 4 % acetic acid, they found the same low amount of migrated silver as in water, n-hexane and in 95 % ethanol even at 50°C. This result is in contrast to our study and findings from other studies (*21, 22*), since in the food simulant acetic acid the migration of silver from polyolefins is expected to be much higher than in water and in organic solvents. The authors concluded that the silver migrated in form of particles because they found the silver of an aqueous migration sample in particular form by scanning electron microscopy (SEM). However, it has to be emphasized that the migration solutions had to be worked up and down-concentrated for SEM analysis first. The concentrated migration sample was first mixed with ethanol and then the solvent was evaporated. Therefore it is not unlikely that the migrated silver ions were reduced by ethanol followed by further aggregation to particles during evaporation of the solvent. Another scientific study on commercial food containers marketed as containing 'nano-' or 'micro-silver' by Goetz et al. (*22*) found similar results for total silver as in our study for migration of total silver. They found the highest migration of silver in 3 % acetic acid food simulant whereas in olive oil no migration was found. They plausibilised the high and fast release of silver with the diffusion of silver ions rather than diffusion of Ag-NPs themselves. Nevertheless, they found Ag-NPs by SEM, transmission electron microscopy (TEM) and single particle inductively coupled plasma – mass spectrometry (SP-ICP-MS). However, the way how this migration test was carried out has to be considered. To increase analytical sensitivity for these measurements to investigate the silver species, the migration test was designed with a high surface to volume ratio of 10 cm² per ml food simulant which corresponds to 10 ml food simulant per 1 dm² contact area: It is obvious that for this high ratio the samples had to be cut into small pieces in which case the so-called cutting edge effect (these containers may have thicknesses of around 1 mm) will have a large impact by generating unrealistic contact conditions. Whereas silver particles on the injection moulded surface are completely covered by the polymer, in cutting edges the particles might be in direct contact with the simulant. This effect increases with increasing nanoparticle concentrations in the polymer. In our experiments we used thin films (60 μm) and the share of cutting edges related to the total surface in the immersion experiment was small. Nevertheless, compared to Ag-NPs at primary particle size of 10 nm in diameter, cutting-edge effects might also play a role. Especially in case of 3 % acetic acid and 10 % ethanol which are simulants with oxidative dissolution potential for Ag-NPs, a diffusion like migration of silver ions is facilitated. The simulants can either attack the Ag-NPs directly or/and penetrate deeper into the polymer at these cutting edges followed by release of silver ions from there. It should be noted that 95% ethanol and iso-octane – both are well-known as aggressive simulants to polyolefins - did not cause any measurable Ag migration which can be considered as a proof that the migrating species was not Ag-NPs.

Due to the inherent analytical difficulties in the unambigous measurement of Ag-NPs in food simulants we extended a migration model which has been applied to usual plastic additives since many years and is generally recognized for this 'conventional' application. However, the extension of the model is still within the logic of the conventional system which is: the mobility or diffusion of a migrant in the polymer is dependent on its size or cross section area. With the selection of spherical NPs consisting of carbon (low atom weight) and having a density of 1 g cm^{-3} low quasi-molecular weights are achieved for the different sphere diameter which means upper limit diffusion coefficients according to equation 2. Therefore these model carbon NPs can be considered as surrogate NPs for any other NP consisting of any chemistry with the consequence that the modeled diffusion coefficients would be universally applicable as a worse case parameter to any other NP of the same size. Consequently, also the modeled migration values (Figure 8) would be represent the worst case for any other NP of the same size.

The modeled migration values according to Figure 8 indicate that only for very small nanoparticles up to 4 or 5 nm in diameter would have a potential to migrate. In the scenario modeled in Figure 8 migration for a 5 nm particle would already be in sub ng kg^{-1} range. However, this would require that concentration of 25000 mg kg^{-1} would consist of 5 nm nanoparticles only which is fully unrealistic and will most likely be never the case in FCM plastics. Since modeled migration decreases then exponentially with bigger diameters migration of nanoparticles into foods appears to be impossible with the consequence that exposure of the consumer to nanoparticles from FCM plastics can be expected to be negligibly low and in any case analytically not measurable. Taking realistic sizes of nanoparticles in FCM plastics into account migration modeling even allows the conclusion that migration will be zero.

Conclusion

The experimental findings of this study show that silver from nano-silver containing polyolefins does only migrate in the ionic silver form (Ag$^+$) in contact with acidic and aqueous food simulants but does not lead to any measurable migration of nanoparticles in any of the official food simulants. The results of a migration model established for nanoparticles in polyolefins even allow the conclusion that not only silver nanoparticles in particular but also nanoparticles in general would not be able to migrate from polyolefins into foods. This conclusion is also based on the fact that typical size distributions of nanoparticles in polymers do in general not contain particles smaller than 5 nm in diameter. Since LDPE is generally recognized as the plastic material with the highest diffusivity for migrants these conclusions are most likely transferable to any type of other plastics. In migration testing for nanoparticles attention must be paid to a correct test procedure avoiding large sample cut edge areas and polymer network destruction by too aggressive test conditions to avoid artefacts and false-positive test results. In any case TEM images should be taken from the test sample to screen for the presence of very small nanoparticles as the only migration-relevant ones.

From the results of this study it can be expected that under normal conditions of use there is no exposure of the consumer to nanoparticles from FCM plastics.

Acknowledgments

We greatfully acknowledge financial support of PlasticsEurope for the comprehensive experimental part of this study. We express our thanks also to the Bavarian State Ministry of Environment and Public Health for financial support of the migration theoretical part of this study which was carried out within the project 'LENA' on 'Nanotechnology related Food Safety' under the coordination of the Bavarian Authority for Public Health and Food Safety (LGL). The authors highly acknowledge analytical-technical support from Gerd Wolz, Fraunhofer IVV.

References

1. United States-Environmental Protection Agency (EPA). *State of Science - Literature Review: Everything Nanosilver and More*; 2010.
2. Wiley, B.; et al. Synthesis of Silver Nanostructures with Controlled Shapes and Properties. *Acc. Chem. Res.* **2007**, *40*, 1067–1076.
3. Azeredo, H. Nanocomposites for food packaging applications. *Food Res. Int.* **2009**, *42* (9), 1240–1253.
4. SCENIHR. *Risk Assessment of Products of Nanotechnologies*; Scientific Committee on Emerging and Newly Identified Health Risks: 2009.
5. Hadrup, N.; et al. Subacute oral toxicity investigation of nanoparticulate and ionic silver in rats. *Arch. Toxicol.* **2012**, *86* (4), 543–551.
6. Tavares, P.; et al. Evaluation of genotoxic effect of silver nanoparticles (Ag-Nps) in vitro and in vivo. *J. Nanopart. Res.* **2012**, *14* (4), 1–7.
7. Das, P.; et al. Effects of silver nanoparticles on bacterial activity in natural waters. *Environ. Toxicol. Chem.* **2012**, *31* (1), 122–30.
8. Lee, Y. J.; et al. Ion-release kinetics and ecotoxicity effects of silver nanoparticles. *Environ. Toxicol. Chem.* **2012**, *31* (1), 155–159.
9. Lok, C. N.; et al. Silver nanoparticles: partial oxidation and antibacterial activities. *J. Biol. Inorg. Chem.* **2007**, *12* (4), 527–534.

10. Xiu, Z. M.; et al. Negligible particle-specific antibacterial activity of silver nanoparticles. *Nano Lett.* **2012**, *12*, 4271–4275.

11. Dallas, P.; et al. Silver polymeric nanocomposites as advanced antimicrobial agents: classification, synthetic paths, applications, and perspectives. *Adv. Colloid Interface Sci.* **2011**, *166* (1–2), 119–135.

12. Chen, X.; et al. Nanosilver: a nanoproduct in medical application. *Toxicol. Lett.* **2008**, *176* (1), 1–12.

13. Buzea, C.; et al. Nanomaterials and nanoparticles: Sources and toxicity. *Biointerphases* **2007**, *2* (4), MR17–MR71.

14. Chaudhry, Q.; et al. Applications and implications of nanotechnologies for the food sector. *Food Addit. Contam., Part A* **2008**, *25* (3), 241–258.

15. Duncan, T. V. Applications of nanotechnology in food packaging and food safety: barrier materials, antimicrobials and sensors. *J. Colloid Interface Sci.* **2011**, *363* (1), 1–24.

16. Silvestre, C.; et al. Food packaging based on polymer nanomaterials. *Prog. Polym. Sci.* **2011**, *36* (12), 1766–1782.

17. Savolainen, K.; et al. Nanotechnologies, engineered nanomaterials and occupational health and safety – A review. *Safety Sci.* **2010**, *48* (8), 957–963.

18. EFSA. *Guidance on the risk assessment of the application of nanoscience and nanotechnologies in the food and feed chain 1*; European Food Safety Authority, 2011.

19. European Commission. Commission Regulation (EU) No 10/2011 of 14 January 2011 on plastic materials and articles intended to come into contact with food. *Official Journal of the European Union* **2011**, *L 12/1*.

20. Cushen, M.; et al. Migration and exposure assessment of silver from a PVC nanocomposite. *Food Chem.* **2013**, *139* (1–4), 389–397.

21. Song, H.; et al. Migration of silver from nanosilver-polyethylene composite packaging into food simulants. *Food Addit. Contam., Part A* **2011**, *28* (12), 1758–1762.

22. Goetz, N.; et al. Migration of silver from commercial plastic food containers and implications for consumer exposure assessment. *Food Addit. Contam., Part A* **2013**.

23. Huang, Y.; et al. Nanosilver Migrated into Food-Simulating Solutions from Commercially Available Food Fresh Containers. *Packag. Technol. Sci.* **2011**, *24* (5), 291–297.

24. Simon, P.; et al. Migration of engineered nanoparticles from polymer packaging to food – a physicochemical view. *J. Food Nutr. Res.* **2008**, *47* (3), 105–113.

25. Crank J. *The mathematics of diffusion*; Oxford University Press: 1975 (ISBN 0-19-853411-6).

26. Simoneau, C., Ed. *Applicability of generally recognised diffusion models for the estimation of specific migration in support of EU Directive 2002/72/EC JRC Scientific and Technical Reports (EUR 24514 EN 2010)*; 2010.

27. Tosa, V.; Mercea, P. Solution of the Diffusion Equation for Multilayer Packaging. In *Plastic Packaging: Interactions with Food and Pharmaceuticals*; Piringer, O. G., Baner, A. L., Eds.; Wiley-VCH: Weinheim, 2008; pp 247–262.

28. Roduit, B.; et al. Application of Finite Element Analysis (FEA) for the simulation of release of additives from multilayer polymeric packaging structures. *Food Addit. Contam.* **2005**, *22* (10), 945–955.

29. Piringer, O. G. A Uniform Model for Prediction of Diffusion Coefficients with Emphasis on Plastic Materials. In *Plastic Packaging: Interactions with Food and Pharmaceuticals*; Piringer, O. G., Baner, A. L., Eds.; Wiley-VCH: Weinheim, 2008; pp 163–193, equation (6.28) on page 188.

30. Zülch, A.; et al. Measurement and modelling of migration from paper and board into foodstuffs and dry food simulants. *Food Addit. Contam.* **2010**, *27* (9), 1306–1324.

31. FABES. *Software MIGRATEST©EXP*; FABES ForschungsGmbH: Munich, Germany, November 2011.

32. AKTS. *Software AKTS-SML*, Version 4.54; AKTS AG, 3960 Siders, Switzerland, 2010.

33. Welle, F. A New Method for the Prediction of Diffusion Coefficients in Poly(ethylene terephthalate). *J. Appl. Polym. Sci.* **2013**, *129*, 1845–1851 (DOI:10.1002/APP.38885).

34. Limm, W.; et al. Modeling of additive diffusion in polyolefins. *Food Addit. Contam.* **1996**, *13*, 949–967.

35. Song, Y.; et al. Static liquid permeation cell method for determining the migration parameters of low molecular weight organic compounds in polyethylene terephthalate. *Food Addit. Contam., Part A* **2013**, *30* (10), 1837–1848.

36. Kittler, S. Synthesis, solubility and biological activity of silver nanoparticles - Ph.D. thesis, in Institute of Inorganic Chemistry, University of Duisburg-Essen, 2009.

37. Alexandrova, V. A.; et al. Silver-carboxymethyl chitin nanocomposites. *Polym. Sci., Ser. A* **2013**, *55* (2), 107–114.

38. Koch, M.; et al. Use of a silver ion selective electrode to assess mechanisms responsible for biological effects of silver nanoparticles. *J. Nanopart. Res.* **2012**, *14* (2), 1–11.

39. von der Kammer, F.; et al. Separation and characterization of nanoparticles in complex food and environmental samples by field-flow fractionation. *TrAC, Trends Anal. Chem.* **2011**, *30* (3), 425–436.

40. Poda, A. R.; et al. Characterization of silver nanoparticles using flow-field flow fractionation interfaced to inductively coupled plasma mass spectrometry. *J. Chromatogr. A* **2011**, *1218* (27), 4219–4225.

41. Bouby, M.; et al. Application of asymmetric flow field-flow fractionation (AsFlFFF) coupled to inductively coupled plasma mass spectrometry (ICPMS) to the quantitative characterization of natural colloids and synthetic nanoparticles. *Anal. Bioanal. Chem.* **2008**, *392* (7-8), 1447–57.

42. Giddings, J. C. Field-Flow Fractionation: Analysis of Macromolecular, Colloidal, and Particulate Materials. *Science* **1993**, *260*, 1456–1463.

43. Shannon, R. D. Revised Effective Ionic Radii and Systematic Studies of Interatomie Distances in Halides and Chaleogenides. *Acta Crystallogr.* **1976**, *A32*, 751–767.

Chapter 6

Food Packaging: Strategies for Rapid Phthalate Screening in Real Time by Ambient Ionization Tandem Mass Spectrometry

Elizabeth Crawford,[*,1] Catharina Crone,[2] Julie Horner,[3] and Brian Musselman[1]

[1]IonSense, Inc., 999 Broadway Suite 404, Saugus, Massachusetts 01906, United States
[2]Thermo Fisher Scientific GmbH, Hanna-Kunath-Strasse 11, Bremen 28199, Germany
[3]Thermo Fisher Scientific, Inc., 355 River Oaks Parkway, San Jose, California 95134, United States
*E-mail: crawford@ionsense.com

Phthalate monitoring in food stuffs is of great interest to monitor for regulated phthalates that could leach into food commodities from packaging and processing. The inherent abundance of phthalates in the environment presents analytical challenges due to the high risk for sample contamination ranging from the equipment used for analytical sample preparation to carryover. Ambient ionization permits ionization of the sample with little to no preparation/sample manipulation and presents the means to directly ionize samples including the packaging, as well as characterize the food commodity itself. The Direct Analysis in Real Time (DART) ambient ionization technique coupled with tandem mass spectrometry offers the ability to rapidly profile phthalates in food packaging and screen for phthalates in consumer products. Collision induced dissociation (CID) fragmentation allows for identification and confirmation for the majority of the phthalates, even for isomeric phthalates based on the relatively gentle fragmentation energies applied. In combination with the higher-energy collisional dissociation (HCD) fragmentation, all phthalates screened for could be differentiated. Characteristic MS/MS fragment patterns in combination with the rapid DART ionization technique and the high scan speed of the latest ion trap mass spectrometer enables a very rapid analytical screening method.

Introduction

Phthalic acid diesters, commonly known as phthalates, are widely used in industry as plasticizers in everyday products like toys, flooring, personal care products, sports equipment and food packaging. In recent years great attention has been paid to the content of phthalates in children's toys, food contact materials and cooking utensils stemming from EU legislations (*1, 2*) and the United States' Environmental Protection Agency's (US EPA) Phthalates Action Plan (*3*) and the United States' Food and Drug Administration's (US FDA) Code of Federal Regulations Title 21 (CFR Title 21) (*4*). The health risks associated with high exposure to phthalates are higher risk of liver and kidney toxicity (*5*), developmental issues in children, increased risk of complications with reproductive health (*6*), in addition to an environmental threat being persistent environmental contaminants, as phthalates are classified as semi-volatile organic compounds (SVOC) and evaporate into the environment over long periods of time.

Phthalates were found in 2011 as being deliberately added to Taiwanese sports drinks to enhance the mouthfeel of the beverages (*7*) and consumers unknowingly were directly ingesting phthalates. Not only were Taiwanese sports beverages affected, but nutraceutical supplements were also found to be contaminated with phthalates. Phthalate content, as well as other additives in food contact materials, such as grease proofers, colorants and stabilizers also should be closely monitored (*8, 9*). With the vast number of food products produced each year and making it onto the consumer market, there is a large demand for rapidly and reliably monitoring for targeted and non-targeted immerging food additives and contaminants.

Typically phthalates are monitored and quantitatively measured by gas chromatography mass spectrometry (GC-MS). This requires labor intensive sample preparation steps for the analyst (*10*), is destructive to the sample, uses large volumes of organic solvents and requires relatively long runs times in comparison with ambient ionization methods where the ionization occurs instantaneously in the absence of chromatographic separation. In 2004-2005, ambient surface ionization was introduced in combination with mass spectrometry in the form of the desorption electrospray ionization (DESI) (*11*) source and the direct analysis in real time (DART) (*12*) ionization source. The development of these instruments enabled the possibility to directly sample from a solid surface and obtain mass spectrometric (MS) data in real time, which meant that the analyst could now completely skip or significantly reduce the sample preparation prior to direct MS data collection. Mass spectrometry is a vital analytical tool for both qualitative screening of known and unknown compounds for structural identification, and in quantitative methods for determining relative levels of targeted contaminants. Fast and robust screening methods are needed in order to reduce the sample burden for the analyst. As a result, the analyst can then redirect their skills and valuable instrument time to the critical samples that need to be subjected to confirmatory analyses employing extensive validated methods. The screening approach utilizing ambient ionization is a great advantage to quickly distinguish regulated substances from the permitted or non-hazardous ones.

The DART ion source used in this work and highlighted in the work at the US FDA by Ackerman *et al.* (*8*, *9*) and Self *et al.* (*7*) and at universities in Germany and Hungary by Rothenbacher *et al.* (*13*, *14*) and Kuki *et al.* (*15*) is demonstrated as a rapid means of direct surface analysis for screening and quantifying phthalates and other additives in food packaging and consumer products. As a targeted screening method coupled with low resolution tandem mass spectrometry (MS/MS) even isomeric phthalates could be distinguished from one another by their MS/MS spectral profiles. Non-targeted screening can be further assessed by coupling ambient ionization with high resolution full scan MS followed by MS/MS for enhanced isomeric differentiation and structural confirmation. In this work, diverse consumer products, including various food packagings, plastic utensils and sports equipment were screened using the developed DART-MS/MS targeted phthalate screening method. Screening of sporting equipment is also of popular interest under the subject of human phthalate exposure, with regards to the transfer of phthalates from direct exposure from the hands to the face, especially to the eyes, which is most common during exercise. A yoga mat and resistance training band were screened under this method. The most commonly observed phthalates found in this study were Di-*n*-octyl phthalate (DOP), Diethylhexyl phthalate (DEHP), Diisodecyl phthalate (DiDP) and Diisononyl phthalate (DiNP), all of which are regulated in the European Union and in the United States in consumer products.

Experimental Set-Up

A targeted tandem MS method was developed with the DART standardized voltage and pressure (SVP) ionization source (IonSense, Inc., Saugus, MA) coupled with the linear ion trap technology in the Velos Pro (Thermo Fisher Scientific, San Jose, CA) mass spectrometer. Twelve phthalate standards (all > 99 % purity) and an EPA mixture of phthalates and adipates (EPA 506 Phthalate Mix, 40077-U analytical standard) were purchased from Sigma-Aldrich (St. Louis, MO) including dimethyl terephthalate (DMTP), diethyl phthalate (DEP), diisopropyl phthalate (DPrP), Di-*n*-butyl phthalate (DBP), Diisobutyl phthalate (DiBP), Di-*n*-propyl phthalate (DPP), Benzyl butyl phthalate (BBP), Bis(2-ethylhexyl) phthalate (DEHP), Di-*n*-octyl phthalate (DOP), Dioctyl terephthalate (DOTP), Diisononyl phthalate (DiNP) and Diisodecyl phthalate (DiDP). For the DART-SVP ion source and MS/MS fragmentation energy optimization experiments, the phthalate standards, all of which were liquids with the exception of DMTP, which was a solid, were directly introduced to the DART-SVP source in their natural state without dilution or dissolution.

Stable and continuous ion signal was generated from the liquid standards by sampling from a single droplet (approximately 3 μL) suspended on the end of a closed-end glass melting point capillary tube. The DMTP solid granules were directly introduced to the DART-SVP source as held by a pair of tweezers mounted onto the automated sampler on the source. The samples were repeated at varying DART-SVP heater temperatures of 150, 250 and 350° C with full scan MS data acquisition in order to determine the optimal desorption temperature with the most abundant ion signals for the $[M+H]^+$ species for the majority of the phthalates. Under the optimized global DART-SVP heater setting of 250° C, the MS/MS collision induced dissociation (CID) activation energies were ramped from 20 – 40 in increments of 5 through the Velos Pro Xcalibur instrument method. The low energy CID fragmentation was not enough to produce confidently distinctive fragment profiles for two isomeric phthalates, DiBP and DBP (nominal m/z 279 for the $[M+H]^+$ ion) and therefore the higher-energy collisional dissociation (HCD) fragmentation option using the minimal energy setting of 10 was employed to yield a higher degree of fragmentation for confident differentiation. The optimized CID and HCD fragmentation values are listed in Table I along with the expected fragment ions and the respective governing regulations (US EPA, US FDA and European Union (EU)). The EPA Mix 506, containing six phthalates DMTP, DEP, DBP, BBP, DEHP and DOP and the adipate bis(2-ethylhexyl) adipate (DEHA), was used to evaluate the optimized MS/MS method before directly analyzing the collected consumer product and packaging samples. The plastic consumer products were collected during normal daily activities and included lid gaskets from metal lids of glass food jars, plastic bags, various plastic packagings, straws, cups, plastic utensils and plastic sports equipment, including a yoga mat and resistance training band.

The DART-SVP and Velos Pro MS were both operated in positive ion mode. The instrument method for the Velos Pro consisted of four scanning events, the first of which contained all of the targeted phthalates with their optimized CID energies in a user defined MS/MS mass list, the second was the targeted HCD optimized scan for DiBP and DBP, the third scan event was a data dependent CID MS/MS scan and the fourth event was a full scan MS over m/z 50 – 500 range. The data dependent scan confirmed the data acquired in the targeted MS/MS scans, as well as to potentially characterize other possible contaminants relative to the full MS scan. Ultra-high purity Helium (99.999 %) and high purity Nitrogen (99.5 %) were used for both the DART-SVP source and Velos Pro mass spectrometer. The DART-SVP source operated exclusively with Helium (80 psi/5.5 bar regulated input) during the sample analysis and switched to Nitrogen (80 psi/5.5 bar input) as a standby gas between sample runs. The capillary temperature on the Velos Pro was set to 200° C and the S lens was held at 50 (arbitrary units) throughout all of the runs. The samples were introduced to the DART-SVP source by holding the objects with a pair of tweezers mounted onto the source and directly positioned into the heated Helium gas beam (Figure 1). The total DART-MS/MS analysis time per sample was less than 30 seconds.

Table I. List of phthalates regulated in EU legislation and under US Environmental Protection Agency's "Phthalates Action Plan" and US FDA's CFR 21. The highlighted phthalates are isomeric and differentiated based on fragmentation.

Compound	Elemental composition	Precursor [M+H]+	Selection of characteristic fragments nominal *m/z*	Optimized Collision Energy	Regulation
Di-n-butyl phthalate (DBP)	$C_{16}H_{22}O_4$	279.2	205.1	HCD 10	US / EU
Diisobutyl phthalate (DiBP)	$C_{16}H_{22}O_4$	279.2	205.1; 223.1, 57.1	HCD 10	US / EU
Benzyl butyl phthalate (BBP)	$C_{19}H_{20}O_4$	313.1	205.1; 239.1; 91.1	CID 40	US / EU
Bis(2-ethylhexyl)phthalate (DEHP)	$C_{24}H_{38}O_4$	391.3	279.2; 261.1; 167.0; 113.1	CID 20	US / EU
Di-n-octyl phthalate (DOP)	$C_{24}H_{38}O_4$	391.3	261.1; 167.0	CID 20	US / EU
Dioctyltere phthalate (DOTP)	$C_{24}H_{38}O_4$	391.3	262.1; 167.0; 111.1; 280.1	CID 20	Not Regulated
Diisononyl phthalate (DiNP)	$C_{26}H_{42}O_4$	419.3	275.2; 127.1; 293.2	CID 25	US / EU
Diisodecyl phthalate (DiDP)	$C_{28}H_{46}O_4$	447.3	289.2; 141.2; 307.2; 99.1	CID 25	US / EU

Data processing was carried out both manually through Xcalibur 2.2 (Thermo Fisher Scientific, San Jose, CA) QualBrowser and automated with MassMoutaineer software developed by Dr. Robert Cody as an MS platform neutral spectral library and search from list program for high resolution and low resolution MS and MS/MS data. The manual data processing was directly compared with the results from the automated processing approach. Predicted fragmentation pathways for the isomeric phthalates DiBP and DBP were generated using MassFrontier software (Thermo Fisher Scientific, San Jose, CA).

Figure 1. The DART-SVP ion source directly interfaced to the Velos Pro linear ion trap mass spectrometer. Direct phthalate screening from the surface of a plastic straw.

Results

Method Development

The DART ionization process is characterized as an atmospheric pressure chemical ionization (APCI) based technique (*12, 16–18*), which generally yields very simple to interpret spectra dominated by [M+H]$^+$ ions in positive ion mode and [M-H]$^-$ ions in negative ion mode. In this study for rapid phthalate screening, all analyses were carried out in the positive ion mode yielding only the [M+H]$^+$ ionic species. The initial analyses that were performed to develop this phthalate screening method were measured on a single stage high resolution mass spectrometer without the possibility for targeted fragmentation. The issue of how to identify the major targeted isomeric phthalates, m/z 279 [M+H]$^+$ for DiBP and DBP; m/z 391 [M+H]$^+$ for DEHP, DOP and DOTP, with the absence of chromatographic separation with DART ionization and not being able to generate selective and optimal fragmentation profiles on other MS systems lent itself to coupling the DART-SVP source with a linear ion trap MS with versatile fragmentation options. This combination allowed the possibility to explore both lower energy collision induced dissociation (CID) and higher-energy collisional dissociation (HCD) fragmentation approaches.

During the MS method development, it was observed that lower energy CID fragmentation was optimal for almost all of the phthalates with the exception of the DiBP and DBP isomeric phthalates. The CID and HCD optimized fragmentation energies shown in the mass spectra in Figure 2 show that using the lowest activation energy setting of 10 for the higher energy HCD fragmentation allowed for two additional ions found at m/z 57 and 167 to be generated for the DiBP fingerprint and only one additional ion at m/z 167 for DBP. The presence of the newly formed fragment ion at m/z 57, in combination with the m/z 223 fragment were used to unambiguously differentiate DiBP from DBP. The CID experiments for DiBP and DBP using the optimized activation energy of 40, yielded only a very small amount of the unique fragment at m/z 223 from the DiBP standard and therefore the differentiation between DiBP and DBP could not be conclusively made using the CID fragmentation approach.

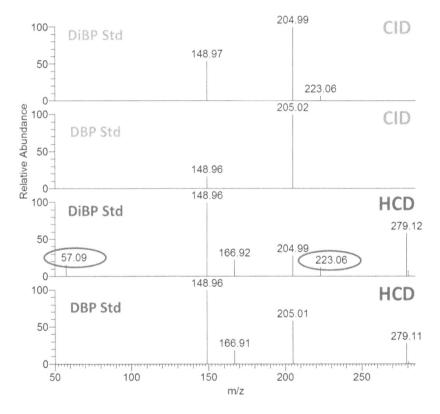

Figure 2. CID and HCD fragmentation fingerprints for isomeric phthalates DiBP and DBP; [M+H]⁺ m/z 279. HCD higher energy fragmentation allows more distinct differentiation.

The other three isomeric phthalates DEHP, DOP and DOTP, of which DEHP and DOP are regulated, all produce the [M+H]⁺ species at nominal m/z 391. Dioctyl terephthalate (DOTP), which is substituted in the para position does not yield the characteristic phthalate fragment ion at *m/z* 149, and generates an MS/MS profile with different relative intensities from both DEHP and DOP. Figure 3 shows the MS/MS spectra for all three *m/z* 391 isomeric phthalates where DOTP stands out from both the DEHP and DOP spectra, and DEHP can be clearly distinguished from DOP based on the presence of the ions at *m/z* 167 and 279. In addition, the relative ion ratios of the fragment ions found at *m/z* 113, 149 and 261 for DEHP and DOP were significantly different, especially for the *m/z* 261 fragment. The predicted ion fragmentation pathways for the closely related DEHP and DOP as proposed by Rothenbacher *et al.* (*13*), were also generated using MassFrontier software, and predict the preferred fragmentation for DOP solely as fragments at *m/z* 149 and 261 (Figure 4). DEHP was predicted to generate fragments at *m/z* 149, 167 and 279 and only minimally at *m/z* 261, which correlated with the measured DART-MS/MS data in Figure 3.

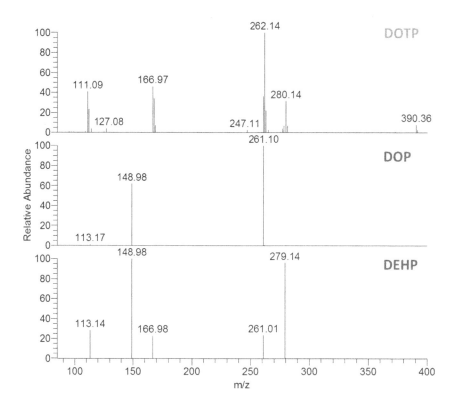

Figure 3. CID MS/MS spectral fingerprints for isomeric phthalates DOTP and the regulated phthalates DOP and DEHP; [M+H]⁺ m/z 391.

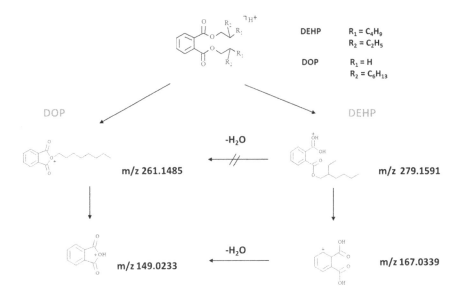

Figure 4. Predicted fragmentation pathways for isomeric phthalates DOP and DEHP using MassFrontier software.

Method Evaluation

The developed DART-MS/MS method was first evaluated using the EPA 506 Phthalate Mix, which included six phthalates and in addition bis(2-ethylhexyl) adipate (DEHA) before applying the method to food packaging and consumer product materials. The standard mixture contained 1:1 ratios (w/w) of all components and was sampled directly from a glass melting point capillary as a liquid (3 µL) without dilution. All of the components were confirmed in the mixture by their MS/MS spectral fingerprints and isomeric DEHP and DOP were both identified. The MS/MS spectra in Figure 5 confirm that DEHP was present based on the presence of the fragment ions at m/z 167 and 279 and that isomeric DOP was also present due to the larger relative abundance of the peak at m/z 261. It was clear from the MS/MS data that there was a mixture of both DEHP and DOP in the standard and the other four phthalates were all positively identified. This result was enough to confidently evaluate the functionality of the DART-MS/MS method. In order to directly determine the exact proportions of the DEHP and DOP in the mixture from the MS/MS spectral profile data, this would require generating calibration curves of varying ratio concentrations of the two phthalates and then rely on matching the spectral fingerprints for a hit.

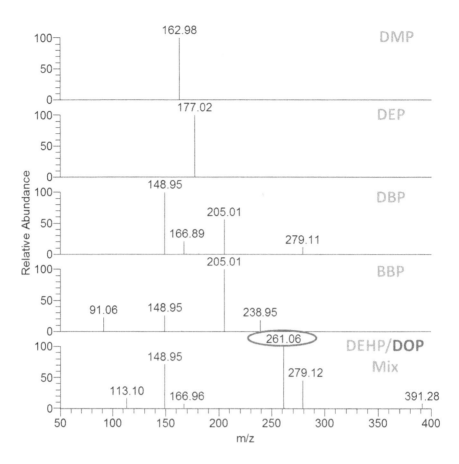

Figure 5. DART-MS/MS method evaluation with EPA 506 Phthalate Mix, confirmation of the six phthalates in the mix, including the mixture of DEHP and DOP, as seen from the MS/MS fingerprint from m/z 391.

Screening Consumer Products and Food Packaging

The results from screening the various food packagings and consumer products with the DART-MS/MS method are summarized in Table II. The most commonly observed regulated phthalates were DEHP, DOP and DiNP. The cases where both DEHP and DOP were present in the sample, in both the plastic knife and fork samples, demonstrate the benefit of the controlled sample introduction to the DART source. In Figure 6, it was observed that DOP was immediately present in the plastic fork and as the fork was held fixed in the heated DART gas beam, DEHP was subsequently ionized as more heat was introduced to the sample and the outermost surface of the fork was desorbed. The spectra show clearly the desorption of only DOP, followed by the mixture of the isomeric DOP and DEHP and then finally only predominately DEHP.

Table II. List of screened consumer products and detected phthalates.

	DMP	DEP	DPrP	DiBP	DBP	DPP	BBP	DEHP	DOP	DiNP	DiDP
Coffee Stick				√				√		√	
Plastic Fork								√	√	√	?[a]
Plastic Spoon								√		√	?[a]
Plastic Knife								√	√	√	?[a]
Croatian Lid									√		
Smoothie Cup								√		√	√
Smoothie Lid							√	√		√	√
Smoothie Straw								√		√	?[a]
Yoga Mat					√			√		√	√
Resistance Band								√		√	

[a] DiDP marked with "?", the relevant ion fragments were present, but the correction ratios were not confirmed.

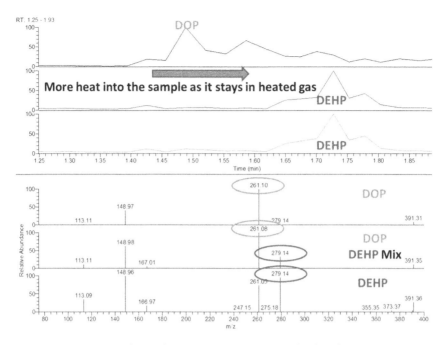

Figure 6. Extracted ion chromatograms (XICs) and related MS/MS spectra sampling directly from an airline plastic fork containing both DOP and DEHP.

Manual data processing was first performed by generating extracted ion chromatograms (XICs) for the major fragment ions of the targeted phthalates and a saved XIC template was directly applied to screen the raw data. More detailed visual inspection of the data then confirmed the presence or absence of the phthalate(s) in the sample. The results reported in Table II for the consumer products were first compiled from the manual data processing while the MassMountaineer software was under development for handling MS/MS spectral data. The same raw data was then reprocessed against the MS/MS spectra for the phthalate standards in MassMountaineer, which confirmed the results from the manual processing. The automated spectral processing required that the DART-MS/MS spectra for the phthalate standards be saved to a library and then the raw data was directly screened again the constructed library of phthalates. The Mass Mountaineer software took into account the presence and absence of ions, as well as the ratio of ion abundances when generating the spectral match scores. Figure 7 shows the typical results score report generated by MassMountaineer where a 100% spectral match yields a score of 1000. In the example of the Croatian lid gasket, the presence of DOP was confirmed with scores above 900 and the isomeric phthalate DEHP received spectral matching scores of less than 500, so it was not considered to be a hit for that sample. Spectral matching scores of 800 and greater were considered to be hits and scores between 500 and 800 were manually inspected for confirmation. The automated processing time per data file was less than 10 seconds compared with the manual data processing, which took on average 2-3 minutes per data file.

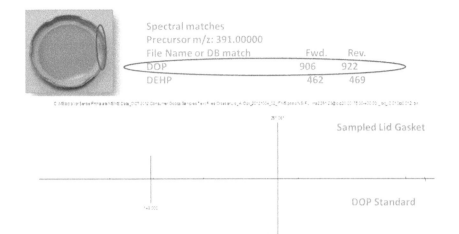

Figure 7. Lid gasket from a Croatian glass food container sampled by DART-MS/MS and data report from the automated MS/MS spectral matching with MassMountaineer software. A 100% match yields a score of 1000.

Conclusions

A rapid phthalate screening method was developed with DART ambient ionization coupled to a low resolution mass spectrometer with tandem MS. The optimized MS/MS method permits real time screening of 12 targeted phthalates, including two sets of isomeric phthalates, and allows the possibility for unknown characterization, for example of other additives that are not in the targeted list based on the data dependent MS/MS scan event. The entire DART-MS/MS analysis time per sample was less than 30 seconds generating spectrally rich MS/MS data. The data processing can be streamlined by using spectral database searching, which simplifies the huge bottleneck of backlogged data interpretation. To demonstrate a rapid screening approach not only does the data generation need to be fast, but the processing and interpretation of results output should match the input. The DART-MS/MS method successfully screened various food packagings and consumer products such as plastic cups, straws and forks, as well as sporting equipment for phthalate content.

Acknowledgments

The authors would like to extend many thanks to Dr. Robert "Chip" Cody for the improvements to the MassMoutaineer software for the scope of this project and his expertise in data processing.

References

1. Petersen, J.; Jensen, L. Phthalates and food-contact materials: enforcing the 2008 European Union plastics legislation. *Food Addit. Contam., Part B* **2010**, *27*, 1608–1616.

2. Fierens, T.; Vanermen, G.; Van Holderbeke, M.; De Henauw, S.; Sioen, I. Effect of cooking at home on the levels of eight phthalates in foods. *Food Chem. Toxicol.* **2012**, *50*, 4428–4435.

3. United States Environmental Protection Agency. Phthalate Action Plan. http://www.epa.gov/opptintr/existingchemicals/pubs/actionplans/phthalates.html (accessed 5 April 2013).

4. United States Food and Drug Administration. CFR Title 21: 178: Indirect Food Additives: Adjuvants, Production Aids, and Sanitizers. http://www.accessdata.fda.gov/scripts/cdrh/cfdocs/cfcfr/CFRSearch.cfm?fr=178.3740&SearchTerm=phthalate (accessed 10 January 2014)

5. Lhuguenot, J. Recent European Food Safety Authority toxicological evaluations of major phthalates used in food contact materials. *Mol. Nutr. Food Res.* **2009**, *53*, 1063–1070.

6. Schettler, T. Human exposure to phthalates via consumer products. *Int. J. Androl.* **2006**, *29*, 134–139.

7. Self, R. L.; Wu, W.-H. Rapid qualitative analysis of phthalates added to food and nutraceutical products by direct analysis in real time/orbitrap mass spectrometry. *Food Control* **2012**, *25*, 13–16.

8. Ackerman, L. K.; Noonan, G. O.; Begley, T. H. Assessing direct analysis in real-time-mass spectrometry (DART-MS) for the rapid identification of additives in food packaging. *Food Addit. Contam., Part A* **2009**, *26*, 1611–1618.

9. Bentayeb, K.; Ackerman, L. K.; Lord, T.; Begley, T. H. Non-visible print set-off of photoinitiators in food packaging: detection by ambient ionisation mass spectrometry. *Food Addit. Contam., Part A* **2013**, *30*, 750–759.

10. Vandenburg, H. J.; Clifford, A. A.; Bartle, K. D.; Carlson, R. E.; Carroll, J.; Newton, I. D. A simple solvent selection method for accelerated solvent extraction of additives from polymers. *Analyst* **1999**, *124*, 1707–1710.

11. Takats, Z.; Wiseman, J. M.; Gologan, B.; Cooks, R. G. Mass spectrometry sampling under ambient conditions with desorption electrospray ionization. *Science* **2004**, *306*, 471–473.

12. Cody, R. B.; Laramée, J. A.; Durst, H. D. Versatile New Ion Source for the Analysis of Materials in Open Air under Ambient Conditions. *Anal. Chem.* **2005**, *77*, 2297–2302.

13. Rothenbacher, T.; Schwack, W. Rapid and nondestructive analysis of phthalic acid esters in toys made of poly(vinyl chloride) by direct analysis in real time single quadrupole mass spectrometry. *Rapid Commun. Mass Spectrom.* **2009**, *23*, 2829–2835.

14. Rothenbacher, T.; Schwack, W. Rapid identification of additives in poly(vinyl chloride) lid gaskets by direct analysis in real time ionisation and single-

quadrupole mass spectrometry. *Rapid Commun. Mass Spectrom.* **2010**, *24*, 21–29.

15. Kuki, A.; Nagy, L.; Zsuga, M.; Kéki, S. Fast identification of phthalic acid esters in poly(vinyl chloride) samples by Direct Analysis In Real Time (DART) tandem mass spectrometry. *Int. J. Mass Spectrom.* **2011**, *303*, 225–228.

16. Song, L.; Dykstra, A. B.; Yao, H.; Bartmess, J. E. Ionization Mechanism of Negative Ion- Direct Analysis in Real Time: A Comparative Study with Negative Ion-Atmospheric Pressure Photoionization. *J. Am. Soc. Mass Spectrom.* **2009**, *20*, 42–50.

17. Song, L.; Gibson, S. C.; Bhandari, D.; Cook, K. D.; Bartmess, J. E. Ionization Mechanism of Positive-Ion Direct Analysis in Real Time: A Transient Microenvironment Concept. *Anal. Chem.* **2009**, *81*, 10080–10088.

18. Gross, J. Direct analysis in real time—a critical review on DART-MS. *Anal. Bioanal. Chem.* **2014**, *406*, 63–80.

Chapter 7

Examination of a Selection of the Patent Medicines and Nostrums at the Henry Ford Museum via Energy Dispersive X-ray Fluorescence Spectrometry

Andrew Diefenbach,[1] Danielle Garshott,[1]
Elizabeth MacDonald,[1] Thomas Sanday,[1] Shelby Maurice,[1]
Mary Fahey,[2] and Mark A. Benvenuto[1,*]

[1]University of Detroit Mercy, Department of Chemistry & Biochemistry,
4001 W. McNichols Rd., Detroit, Michigan 48221
[2]The Henry Ford, 20900 Oakwood Boulevard, Dearborn,
Michigan 48124-5029
*E-mail: benvenma@udmercy.edu.

A series of patent medicines from the collections of the Henry Ford Museum in Dearborn, Michigan were analysed via energy dispersive X-ray fluorescence spectrometry for the presence of a wide range of elements. A series of lead standards as high as 1,000 ppm were produced and used as reference points so that the analyses were not purely qualitative. The presence of heavy metals was found in several of the samples. Less expected was the presence, in numerous samples, of several elements now known to be beneficial to human health, including potassium, calcium, and iron, marking them possibly as an early type of food supplement.

Introduction

For millennia, humankind has utilized plant and animal sources for the relief of pain and the cure of a wide variety of ailments (*1*). As an example, one common pain reliever still used extensively today is aspirin. Originally isolated from the bark of willow trees, early sufferers found that if the tree bark was chewed, pain was lessened. Since such times, aspirin production has been expanded enormously, and in lieu of using plant sources, the starting material for its production today is phenol, usually refined from crude oil.

As the nineteenth century ended and the twentieth began, while a systematic understanding of drug use had begun – for pain relief as well as for other health benefits – unregulated medications and nostrums were routinely manufactured and produced for direct sale to the public, often by individuals who claimed the title "Doctor," as evidenced in Figures 1–3. All the medicines in this study were manufactured without the in vitro, animal, and human testing and oversight we now expect however, in a time before any governmental oversight, such as the Biologics Control Act of 1902 (*1–3*), or the establishment of the Food and Drug Administration or other such governmental agencies (*4–8*). These medicines and nostrums were manufactured in a time when ingredients were generally kept secret (*9, 10*) so that competitors would be unable to steal recipes and replicate a successful product (*11, 12*). Since the producers of such medicines and nostrums have passed away decades ago, and no proof exists that any modern recipes or existing ingredient lists for these medicines are the same as those made at the turn of the twentieth century, an examination of several of these materials which have been stored in the Henry Ford Museum has been undertaken, to attempt to determine their elemental compositions.

What can be learned from examining a group of what are now derogatorily termed 'quack medicines' and 'snake oil?' As background, the packaging on each material does sometimes state the medicine's primary use. Figure 1 for example, a photograph of a box of Dr. Sawen's Magic Nervine Pills, indicates that the manufacturer marketed it as a "nerve vitalizer." Additionally, newspapers of the time, and pamphlets produced by the manufacturers of the medicines (*13*) make claims about what a specific medication can do (sometimes very colorful claims) and what ailments the medicine was meant to treat. But since neither lists ingredients in a manner dictated by law today (*4–6*), and since few previous studies of older patent medicines and nostrums, or any medicinal materials, appear to have utilized X-ray techniques (*14–22*), it was felt that examination via energy dispersive X-ray fluorescence spectroscopy might be an efficient way to determine the presence of a wide variety of elements within these materials, and result in an enhanced understanding of the composition of these medications, beyond what has been established (*22–25*).

Experimental

All samples were analyzed at least three times using a Spectrace QuanX EDXRF spectrometer, which utilizes fundamental parameters software, and pure element standards. Sample excitation conditions were as follows: 20kV, 0.14 mA, 100 sec count, $K_{\alpha\beta}$, palladium medium filter, mid Zb conditions, for Fe, Ni, Cu, Zn, Au, Pb, Bi, As and Co, and 41kV, 0.24 mA, 100 sec count, copper thin filter, high Za conditions for Ag, Sn and Sb. The spectrometer uses a rhodium target X-ray tube, and certified copper and lead samples were run each day prior to sample runs to verify both instrument accuracy and precision.

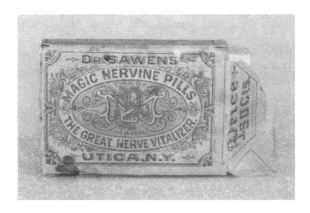

Figure 1. Photograph of Box for Dr. Sawen's Magic Nervine Pills

In order to fully examine each of these samples by X-ray fluorescence, without any bias for what elements might be present, a protocol was developed in order to establish results that were not merely qualitative. A 1,000 ppm solution of soluble lead was run prior to the samples, and comparisons were established.

Certified AAS standard solutions were used: lead (1000 ppm) and its dilutions. Eight standards were prepared for the metal at varying concentrations: 5, 20, 50, 100, 250, 500, 750 and 1000 ppm. A 30 mL sample of each concentration was made by dilution with distilled water. Each standard was stirred vigorously for 15 minutes to ensure homogenous dilution. A 5 mL aliquot of each standard was then transferred into an XRF sample container. The calibration parameters were set as follows: linear analysis, generation of conditions, concentrations, uncertainty, peak intensities, and background intensities (*26*).

The samples were all solid materials, and were all either visually homogeneous powdered material, or were homogenized using an agate mortar and pestle. This ensured that there were no irregularities in the material when they were placed in samples cups for analysis.

Results and Discussion

Figure 2 illustrates each patent medicine examined in the study, at 20kV X-ray intensity, and lists the elements of atomic number 11 or higher (mass 23 or greater) that was found in each. The lead standards that had been produced were used to establish peak height comparisons with other elements, and a numeric value was assigned to each, in an attempt to begin to quantify what would otherwise have been a strictly qualitative examination. Thus, a value of 1 is greater in height than that registered by the 1,000 ppm lead standard, 2 equals the lead 1,000 ppm standard, 3 is half the height, and 4 is one-fourth the height of the 1,000 ppm standard. The value of 5 represents a detectable amount of the element, larger than the limit of detection of the instrument, but less than one-fourth of the 1,000 ppm lead standard peak height.

Sample #	Name	S	Cl	K	Ca	Ti	Fe	Cu	Zn	As	Br	Sr	Ag	Ba	Au	Hg	Pb
HF_93.0.3.90	Dr. Sawen's Magic Nervine Pills			5	3		4	5			4	3		5		4	5
HF_91.0.11.171	Doct. Ingoldsby's Piles Specific			4	3		3	5				4	5				
HF_443	Hollister's Golden Nugget Tablets				2							4	5	5			
HF_2003.0.16.90	Tripeptine Tablets	5			1		4			4		3	5	5			
HF_2003.0.16.149	Lorraine's Vegetable Cathartic Pills			5	5		4	5	5				5		5		4
HF_2003.0.16.272	Diamond Dinner Pills	5		5	5		5	5	5								5
HF_91.0.11.80.7	Dr. Page's Rail Road Pills	5		5	5		4	5						5			3
HF_2003.0.16.292	Dr. Comfort's Candy-covered Cathartic Compills			5	5		5	5				4			5	5	
HF_2003.0.16.166	Eilert's Day Light Family Liver Pills	5		5	5		5	5			5		5			1	
HF_91.0.11.159	P.P.P. Parson's Purgative Pills	5		5	5		5				5		5			2	
HF_2003.0.16.175	Dr. Tutt's Liver Pills	5		5	5		5	5			5		5			1	
HF_2003.0.16.123	Anti Bilious Purgative Pills (laxative)	5	5	5	5	5	5				5		5			2	
HF_2003.0.16.231	Sheldon's Dyspepsia Troches			5	5		1	5	5		5		5				
HF_2003.0.16.259	Dr. Derby's Liveroid Cathartics for Liver Complaints			5	5		2	5	5		5		5				
HF_334_Anti Malaria	Rhodes' Fever & Ague Cure or, Antidote to Malaria						2										
HF_91.0.11.175	Doct. F.G. Johnson's French Female Pills	5		5	5		1	5	5				5				5
HF_91.0.11.2694.3	Dr. J.J. Gallup's Vegetable Family Pills	5		5	4		3	5		5		4	5				5
HF_91.0.11.2699	Reynolds & Parmely's Female Health Restorative	5			5		2	5					5				
HF_00.4.6101.2	Ayer's Cathartic Pills	5		5	5		4	5	5								
HF_2003.0.16.170	Knill's Anti-Dyspepsia Tablets "Pepsin Compound"		5	5	5		4	5	5		5		5	5	5		
HF_2003.0.16.154	Dr. Filkin's Vegetable Sugar Coated Liver Pills			5	5		5	5									5
HF_37.314.2.20	Newton's Jaundice Bitters						5						5				
HF_2003.0.16.310	Dr. Bennet's Plant & Roots Pills	5		5	4		4	5						5			5
HF_2003.0.16.176	DeBell's Kidney Pills			5									5	5	5		
HF_00.4.6116.2	Dr. Freeman's C.D.Q.		5		5		5						5	5	5		

Figure 2. Samples, Elements, Relative Amounts at 20keV

90

While each medicine was produced independently of all the others, and indeed may have been produced as a competitor with some of the others in the study (*9–12*), comparisons between the elements and their relative amounts can prove quite useful. As well, tabulating the data, as has been done in Figure 2, clarifies what common elements are present, and further establishes which patent medicines and nostrums have ingredients that are unique, or at least not common to most of those in this set. As well, the tabulated data provides an indicator of how much of a specific element is in each sample.

Deleterious Elements

As mentioned in the introduction section, there appears to be a common belief today that patent medicines produced before the existence of the US Food and Drug Administration (*2–6*) were usually harmful, or at least were no better than placebos (*7, 9, 10*). It is evident from Figure 2 that several of the medicines examined here did contain lead and mercury. Two did contain arsenic.

Figure 3 lists only those elements that contained detectable amounts of lead, mercury, and/or arsenic. Curiously, while mercury only occurs in 5 of the 25 samples, it does so in relatively high amounts in 4 of those 5. Those four are all advertised either as purgative or liver pills, and indeed, small amounts of mercury were used in laxatives in the past, although such materials have since been phased out.

Lead is present in 10 of the 25 samples in this study, but in only one does it rise significantly above the detection limit. In no case can lead be considered beneficial to human health, and in the sample with the highest amount of lead, Dr. Page's Rail Road Pills, the name of the medicine does not provide any clues as to the reason lead was necessary in the formulation.

Arsenic was present in only two samples, the Tripeptine Tablets and in Dr. J.J. Gallup's Vegetable Family Pills. Like lead, arsenic is never beneficial to human health, although not all oxidation states of arsenic are deleterious. Because there are only two samples which manifest arsenic, and because based on the medicines' names these two samples do not immediately fall into an obvious category, it is difficult to establish why this element was present.

Beneficial Elements

While the presence of materials containing lead, mercury and arsenic is, disappointingly, not unexpected in such antiquated patent medicines, the presence of elements that are now considered quite beneficial – such as potassium, calcium, iron, copper, zinc, and even silver – was unexpected, and becomes an indicator that some of these medicines may indeed have met some of their claims. These are discussed further, below.

Sample #	Name	S	Cl	K	Ca	Ti	Fe	Cu	Zn	As	Br	Sr	Ag	Ba	Au	Hg	Pb
HF_93.3.90	Dr. Sawen's Magic Nervine Pills			5	3		4	5		4	4	3		5		4	5
HF_2003.0.16.90	Tripeptine Tablets	5			1		4		5			3	5	5			4
HF_2003.0.16.149	Lorraine's Vegetable Cathartic Pills			5	5		4	5	5				5		5		4
HF_2003.0.16.272	Diamond Dinner Pills	5		5	5		5	5	5						5		5
HF_91.0.11.80.7	Dr. Page's Rail Road Pills	5		5	5		4	5						5			3
HF_2003.0.16.292	Dr. Comfort's Candy-covered Cathartic Compills			5	5		5	5				4	4			5	
HF_2003.0.16.166	Eilert's Day Light Family Liver Pills			5	5		5	5			5	5	5			1	
HF_91.0.11.159	P.P.P. Parson's Purgative Pills	5		5	5		5				5	5	5			2	5
HF_2003.0.16.175	Dr. Tutt's Liver Pills	5		5	5		5	5			5	5	5			1	5
HF_2003.0.16.123	Anti Bilious Purgative Pills (laxative)	5	5	5	5	5	5	5			5					2	5
HF_2003.0.16.231	Sheldon's Dyspepsia Troches			5	5		1	5	5			5	5				5
HF_91.0.11.175	Doct. F.G. Johnson's French Female Pills			5	5		1	5	5			5	5				5
HF_91.0.11.2694.3	Dr. J.J. Gallup's Vegetable Family Pills	5		5	4		3	5		5		4	5				5
HF_2003.0.16.154	Dr. Filkin's Vegetable Sugar Coated Liver Pills			5	5		5	5				5					5
HF_2003.0.16.310	Dr. Bennet's Plant & Roots Pills	5		5	4		4	5	5					5			5

Figure 3. Samples Containing Lead, Mercury, and / or Arsenic

Common Elements: S, K, Ca, Fe, Cu, Zn, Ag, Pb

An examination of Figure 2 illustrates that these eight elements are present in a large number of the samples. Sulfur is present in small amounts in ten of the 25 samples. Potassium is present in 18 of the 25 samples. Calcium is present in 22 of the 25 samples, and in four cases is present in intensities equal to or nearly equal to the intensity of a 1,000 ppm lead standard. Iron is also present in 23 out of 25 samples, seven of which show high intensities. Copper is present in 15 of the 25 samples, but is seen at low intensity. Silver is also present in 16 of the 25 samples, again at low intensities. Lead, which has been discussed, above, is present in 10 of the 25 samples.

Concerning sulfur: small amounts of sulfur are necessary in the human diet, and can be obtained through certain foods. Plant growth requires a certain amount of sulfur, and since it can be assumed that several of these medicines utilized some plant source as an ingredient, it is not surprising that medicines such as Dr. J.J. Gallup's Vegetable Family Pills would contain it. Some of the medicines examined, such as Dr. Page's Rail Road Pills, do not provide any hints as to why sulfur is present in them.

In regards to potassium, once again, potassium is required in the human diet, and the now common three number designator for plant fertilizers always lists the percentage of potassium as the third number (for example, 10-6-4 fertilizer contains 10% nitrogen, 6% phosphorus, and 4% potassium). Nevertheless, it was unexpected to find potassium in so many of the samples. The ubiquitousness of the element prompted a re-examination of these peaks to ensure there was no possibility of overlap between any other elemental Kα or Kβ lines and those of potassium that could yield a false positive. None were found.

Calcium was found in 22 of the 25 samples, a number exceeded only by iron, and even then, only by one sample. Calcium is certainly essential to the human diet, and is found in a variety of natural sources, from which the medicines were most likely produced. Additionally, calcium compounds have been utilized as antacids for over a century, and the sample with the highest calcium reading, Tripeptine Tablets, might have functioned as an antacid. The name of the patent medicine could be a corruption or deliberate re-wording of "pepsin," a term first used in 1836 in relation to digestion (*27*). Some calcium antacids are made from calcium carbonate, thus the anion would not be visible to the X-ray fluorescence spectrometer.

Iron is the element found in all but two of the 25 samples studied. In the 7 samples in which its intensity was the highest, two have titles indicationg they were targeted towards women, and in maintaining women's health. It is now known that some women do suffer from anemia caused by blood loss during the menstrual cycle. Thus, Doct. F.F. Johnson's French Female Pills, and Reynolds & Parmely's Female Health Restorative may have been attempts to remedy this problem. Medicines such as Sheldon's Dyspepsia Troches may also have been used to such an end, since dyspepsia – a form of indigestion with symptoms that include heartburn and pain – and the pains associated with the menstrual cycle may have been confused, at least by the manufacturer.

Copper is a necessary ionic nutrient in human health, and is present in a variety of animal and vegetable sources. Copper compounds play several different roles in living organisms, and the presence of it in 15 of 25 samples, albeit in relatively small amounts, might be an indicator that it was obtained from different plant or animal sources. Its addition through some inorganic copper source, such as a copper salt, can also not be ruled out, however.

Zinc plays a vital role in human health, and indeed is marketed in multi-vitamins today. The presence of zinc in 9 of the 25 samples here may indicate that the manfacturers knew of the benfits of zinc and zinc-containing materials in aiding human health.

The presence of silver in 16 of the 25 samples was unexpected, but the presence of both copper and silver in 8 of the 16 silver-containing samples has an explanation that today might seem far removed from medical formulations. Silver can today be produced as a by-product of copper refining, because both silver and gold occur naturally in many copper deposits, and copper mineral deposits (28). It may be that the source material for copper in several of these medicines also provided small amounts of silver.

Lead has been discussed in terms of lead compounds being deleterious to human health, but the presence of lead in 10 of the 25 samples here, and in samples that can not be conveniently grouped into one type of medicine – such as purgatives or fever reducers – suggests that lead may be present for another, less obvious reason. Since lead was found to be present in these samples amounts that were small but above the instrument's level of detection, the possibility exists that some lead compounds were included in these medicine recipes simply because such compounds possess a sweet taste. What is sometimes called 'lead white,' or 'litharge,' terms that have been used interchangeably throughout history, is actually lead (IV) oxide, and in very small amounts may have been used to make a medicine formula more palatable, or at least less bitter. It was not possible, in the present investigation, to determine whether or not the amounts of lead detected were immediately toxic to the user. It can be surmised that over the course of time, a person who used such a medicine repeatedly would eventually develop symptoms of lead poisoning.

Other Elements: Strontium, Barium, Gold

The presence of strontium in 5 of the 25 samples, and in samples in which calcium showed an equally strong signal or a stronger signal indicates that the strontium was an impurity in the calcium sources. While strontium can today be directly refined from minerals such as celestite, it does occur naturally in 1% - 2% abundance within many calcium-containing ores.

Barium occurs in 6 of the 25 samples, and like strontium, may be present in these samples because it was present as a minor component of some calcium-containing mineral source.

Gold is only present in 3 of the 25 samples, but in two of them co-exists with copper. As mentioned for silver, above, gold can be recovered from copper refining operations, and can at times be a second product that generates a profit. The amount of gold differs from one source or ore batch to another, and from one location to another. The gold present in two samples here that also contain copper may be cases in which some very small amount of gold was present in the original copper source material. The source of the gold in the third sample, Dr. Freeman's C.D.Q., may be the silver that is also in that sample.

Conclusions

The 25 medicines that were examined in this study displayed a wide variety of elements, many of which have been found in past decades to be beneficial to human health, and some of which are still sold in over-the-counter vitamins, medications, and homeopathic medicines (29).

Those samples that contained what are generally considered harmful elements, including mercury, lead, and arsenic, appear in several cases to be used in laxative or purgative medications. Mercury has played a role in this aspect of human health in the past, though it has been removed from modern products. Very small amounts of lead may have been added to some of the medicines to make them taste better to the user and consumer. Arsenic appears in only two of the samples.

There was a much larger number of samples that contained multiple beneficial elements than was expected. Although the term 'snake oil,' often used for medications like those studied here, has become synonymous with charlatans and medicines that had no positive effects, and that may have had negative effects, it appears that there were cases in which these medicines contained numerous elements that have since been proven to be beneficial to the human diet, such as potassium, calcium, and iron. These twenty five samples thus appear to represent a class of medicines or food supplements that would today be defined as unregulated, but that may have had positive effects on those who used them.

References

1. Cowen, D. L., Helfand, W. H. *Pharmacy, An Illustrated History*; Harry N. Abrams, Inc.: New York, NY, 1990.
2. Kondrates, R. A. In *The Early Years of Federal Food and Drug Control*; Young, J. H., Ed.; American Institute of the History of Pharmacy: Madison, WI, 1982; pp 8–27.
3. *The Propaganda For Reform In Proprietary Medicines*, 5th ed.; American Medical Association, Council On Pharmacy and Chemistry: American Medical Association Press, Chicago, IL, 1908. http://books.google.com/books?id=ZX6OSPD_J8UC&printsec=frontcover&source=gbs_ge_summary_r&cad=0#v=onepage&q&f=false (accessed August 23, 2103).
4. Formation of the Food & Drug Administration. http://www.fda.gov (accessed September 30, 2013).
5. Federal Food, Drug, and Cosmetic Act of Public. Public Law 75-717, 1938.
6. Cowen, D. L. America's First Pharmacy Laws. *J. Am. Pharm. Assoc., Pract. Pharm.* **1942**, *3*, 162–169.

7. Cowen, D. L. The Foundations of Pharmacy in the United States. *J. Am. Med. Assoc.* **1976**, *236*, 83–87.

8. Harden, V. A. *Inventing the NIH: Federal Biomedical Research policy 1887 – 1937*; Johns Hopkins University Press: Baltimore, MD, 1986.

9. Berman, A. The Botanic Practitioners of 19-th Century America. *Am. Prof. Pharm.* **1957**, *23*, 868–70.

10. Berman, A. A striving for Scientific Respectability: Some American Botanics and the Nineteenth-Century Plant Materia Medica. *Bull. Hist. Med.* **1956**, *30*, 7–31.

11. *Facts Worth Knowing: Falsehoods Exposed, The Truth About Patent Medicines, Mercenary And Selfish Character of Attack On Popular Household Remedies by Yellow Journals And Doctors' Organizations*; The Proprietary Association: 1908. http://books.google.com/books?id=0tuG2-pNQkEC&printsec=frontcover&source=gbs_ge_summary_r&cad=0#v=onepage&q&f=false (accessed August 23, 2013).

12. Young, J. H. Patent Medicines: An Early Example of Competitive Marketing. *J. Econ. Hist.* **1960**, *30*, 648–656.

13. *Hostetters Illustrated United States Almanac 1877*; Hostetter & Smith: Pittsburgh, PA, 1877.

14. Zhong, X.-K.; Li, D.-C.; Jiang, J.-G. Identification and Quality Control of Chinese Medicine Based on the Fingerprint Techniques. *Curr. Med. Chem.* **2009**, *16*, 3064–3075.

15. Zheng, X.; Zhong, J.; Feng, Y.; Lv, Y. Identification On Chinese Traditional Medicine Sanhuang Tablets by X-ray Diffraction Fourier Fingerprint Pattern. *Zhongguo Yaoshi* **2007**, *21*, 813–815.

16. Guan, Y.; Ding, X.; Wang, W.; Di, L.; Wang, X. The Methods of X-ray Fluorescent Analysis and X-ray Diffraction of Flos Trollii. *Yaowu Fenxi Zazhi* **2006**, *26*, 1623–1625.

17. Wang, W.; Zhang, H.; Li, X. Determination of Elements of Flos Trollii From Different Producing Areas. *Guangdong Weiliang Yuansu Kexue* **2007**, *14*, 36–37.

18. Wang, W.-J.; Guan, Y.; Zhu, Y.-Y. Novel Identification of Donkeyhide Glue By X-ray Fluorescence Analysis. *Guangpuxue Yu Guangpu Fenxi* **2007**, *27*, 1866–1868.

19. Qiu, H. New Application of X-ray Diffraction in the Identification of Chinese Medicine. *Zhongguo Yaoye* **2005**, *14*, 89–90.

20. Chen, H.; Wu, Y.; Zheng, Q.; Lu, Y. X-ray Diffraction Fourier Fingerprint Pattern Qingkailing Patent Medicine. *Zhongguo Taoke Daxue Xuebao* **2004**, *35*, 138–140.

21. Zheng, X.; Lu, Y.; Zhao, B.; Lin, R.; Zheng, Q. Studies on X-ray Diffraction Fourier Pattern. *Yaowu Fenxi Zazhi* **2000**, *20*, 202–205.

22. Takano, I.; Seto, T.; Yasuda, I.; Hamano, T.; Takahashi, N.; Watanabe, Y. Analysis of Mercury and Arsenic in the Chinese Patent Medicine "Bezoar Antifebrile Pills". *Shoyakugaku Zasshi* **1993**, *47*, 70–73.

23. Griffenhagen, G. B.; Young, J. H. Old English Patent Medicines in America. *Bull. - U.S. Natl. Mus.* **1959**, *218*, 153–183.

24. Helfand, W. H. Ephemera of the American Medicine Show. *Pharm. In Hist.* **1985**, *27*, 185–191.

25. *Balm of America, Patent Medicine Collection: Introduction*; National Museum of American History. http://americanhistory.si.edu/collections/object-groups/balm-of-america-patent-medicine-collection (accessed September 30, 2013).

26. Garshott, D. M.; Macdonald, E. A.; Murray, M. N.; Benvenuto, M. A.; Roberts-Kirchhoff, E. Elemental Analysis of a Variety of Dried, Powdered, Kelp Food Supplements for the Presence of Heavy Metals via Energy-Dispersive X-Ray Fluorescence Spectrometry. In *It's All In the Water: Studies of Materials and Conditions in Fresh and Salt Water Bodies*; Benvenuto, M. A., Roberts-Kirchhoff, E. S., Murray, M. N., Garshott, D. M., Eds.; ACS Symposium Series; American Chemical Society: Washington, D.C., 2011; Vol. 1086, pp 123–133.

27. Fruton, J. S. A history of pepsin and related enzymes. *Q. Rev. Biol.* **2002**, *77* (2), 127–147.

28. Benvenuto, M. A. *Industrial Chemistry*; DeGruyter Publishing: Berlin, Germany, 2013.

29. Coulter, H. L. *Homeopathic Influences in Nineteenth-Century Allopathic Therapeutics*; American Institue of Homeopathy: Washington, DC, 1973.

Chapter 8

Analysis of Nine Edible Clay Supplements with a Handheld XRF Analyzer

Jessica L. LaBond, Nicholas H. Stroeters, Mark A. Benvenuto, and Elizabeth S. Roberts-Kirchhoff[*]

Department of Chemistry and Biochemistry, University of Detroit Mercy, Detroit, Michigan 48221
[*]E-mail: robkires@udmercy.edu

Nine edible clay powder supplements and three soil National Institute of Standards and Technology (NIST) standard reference materials (SRMs) were analyzed for various elements with a handheld X-ray fluorescence (XRF) analyzer using a simple, rapid, and low-cost method. The results from the analysis of the soil SRMs were within 13% of the reported values for iron, titanium, lead, manganese, zinc, and strontium and 20% for copper and calcium. The clay samples were obtained from a variety of locations around the world as advertised by the different suppliers. Each of the clay samples had a different composition as determined by XRF analysis. All of the clay samples had detectable levels of lead and arsenic. The clay samples did not have detectable amounts of cadmium or mercury.

Introduction

Clays consisting of calcium bentonite, calcium montmorillonite or sodium bentonite have been used for cosmetic purposes and as dietary supplements. As dietary supplements, the clays are advertised for use as a mineral source or as a detoxifying agent. Some claims include that the clay contains "all natural" properties known for therapeutic healing or as a classic American intestinal cleanser studied by doctors for its potential use to cure stomach illnesses (*1, 2*). In addition, studies have shown that repeated intake of clays can lead to a disorder known as geophagia; the purposeful consumption of earthy non-food items (*3, 4*).

In 1994, the Federal Food Drug, and Cosmetic Act (5) was amended with the passage of the Dietary Supplement Health and Education Act of 1994 (DSHEA) (6). This law defined a dietary supplement to mean a product (other than tobacco) that, among other things, is intended for ingestion that contains one or more of the following dietary ingredients: vitamins; minerals; herbs or other botanicals; amino acids; dietary substances to supplement the diet by increasing the total daily intake; or concentrates, metabolites, constituents, extracts, or combinations of these ingredients; and is labeled as a "dietary supplement" (6). Dietary supplements and foods containing added dietary ingredients, such as vitamins and herbs, constitute a growing multibillion dollar industry. Sales of dietary supplements alone reached approximately $23.7 billion in 2007, and data from the 2007 National Health Interview Survey show that over half of all U.S. adults consume dietary supplements (7). In 1994, there were approximately 4,000 dietary supplement products on the market, whereas an industry source estimated that, in 2008, about 75,000 dietary supplement products were available to consumers (7). The increasing popularity and use of dietary supplements and the regulations governing this segment of the market has prompted numerous investigations into the quality and purity of these supplements.

Under the DSHEA, the dietary supplement manufacturer is responsible for ensuring that a dietary supplement is safe before it is marketed. This act does not specify constituents of concern or methods for analyses of these. The FDA is responsible for taking action against any unsafe dietary supplement product after it reaches the market. Generally, manufacturers do not need to register their products with the FDA nor get FDA approval before producing or selling dietary supplements (5, 6). In addition, "unlike drug products, manufacturers and distributors of dietary supplements are not currently required by law to record, investigate or forward to FDA any reports they receive of injuries or illnesses that may be related to the use of their products. Under DSHEA, once the product is marketed, FDA has the responsibility for showing that a dietary supplement is "unsafe," before it can take action to restrict the product's use or removal from the marketplace" (5).

It is of interest to study the chemical composition of these clay supplements since they are mined from various locations and refined to various degrees. It is of particular interest to study the presence of heavy metals in these clay supplements. Arsenic, mercury, cadmium, and lead are of primary concern because of their toxicity and the potential to be present as contaminants. Previous studies on the prevalence of these elements in dietary supplements indicate that relatively high concentrations of these elements may occur (8–16). According to the US Pharmacopeia, permitted daily exposure limits in dietary supplements for inorganic arsenic is 15 µg/day, for cadmium is 5 µg/day, for lead is 10 µg/day, and mercury (total) is 2 µg/day (17). The clinical manifestations of chronic arsenic toxicity include peripheral neuropathies, cognitive deficits, fatigue, gastrointestinal complaints and skin afflictions (18). Today, the most common sources of lead exposure in the United States are lead-based paint in older homes, contaminated soil, household dust, drinking water, lead crystal, and lead-glazed pottery. Chronic lead exposure in adults can result in increased blood pressure, decreased fertility, cataracts, nerve disorders, muscle and joint pain, and memory or concentration problems (*19*). Mercury may be released naturally into the air from volcanoes and the earth's crust, but man-made sources include the incineration of waste and coal-burning power plants. Once these industrial activities release mercury into the air, it ultimately falls back to earth, is fixed by plankton into methyl mercury and is concentrated up the food chain by the fish that eat them. Mercury poisoning may include the following symptoms: muscle weakness, skin rashes, mental disturbances such as mood swings and memory loss, impairment of speech, hearing and peripheral vision, impairment of coordinated movements such as walking or writing, and numbness and "pins and needles" feeling in the hands, feet, and sometimes around the mouth (*20*).

The use of handheld X-ray fluorescence (XRF) analyzers is a non-destructive, rapid, and low-cost method to analyze a large number of samples with little sample preparation (*21*). Handheld XRF analyzers have been used to screen for certain elements in soil, building materials, biological materials, food, beverages, and dietary supplements (*22–28*). In this study, nine clay supplements, advertised for use as dietary supplements, were investigated with a handheld XRF analyzer using a soil analysis method that analyzes for various elements including calcium, titanium, manganese, iron, copper, zinc, arsenic, strontium, zirconium, cadmium, cerium, mercury, and lead. National Institute of Standards and Technology (NIST) standard reference materials (SRMs) were analyzed as positive controls. The amounts of arsenic and lead ingested daily were calculated using the concentration, unit dose weight and recommended dosages. These were compared to the *US Pharmacopeia* (USP) specifications for the permitted daily exposure levels for elemental impurities (*17*).

Experimental Methods

Standard reference materials (SRMs), including San Joaquin soil SRM 2709 and two soil standards containing lead paint, SRM 2586 and SRM 2587, were from the National Institute of Standards and Technology (NIST). The soil standards and clay samples were analyzed with a Bruker S_1 TURBO handheld XRF analyzer with a silicon drift detector and a resolution of approximately 145 eV at 200000 cps. Five samples per soil standard (0.80 g each sample) were analyzed five times each for 120 s using the soil FP calibration method. This calibration has been optimized for a SiO_2 matrix, is non-normalized and relies on using a repeatable sample geometry and distance between the sample and detector. All samples were analyzed using a repeatable distance by placing the sample cups on the safety platform that fits on the nose of the instrument. The instrument was mounted in the bench top stand. The voltage and current were 45 kV and 30 µA, respectively, with a Ti/Al filter. The instrument was controlled with a hand held Hewlett Packard PDA with XBruker Elemental S1 software. The data files were imported into Microsoft Excel for statistical analysis. The soil samples were used as provided and placed into Chemplex Spectrocertified® Quality XRF micro-sample cups (31.0 mm x 22.4 mm) and covered with Chemplex Spectromembrane® perforated thin film mylar polyester sample support carrier films (3.6 µm) with a sealing ring. The concentrations (averages and standard deviations) of various elements for the three standards were determined and then compared to the NIST standard values (*30–32*). The Limit of Detection (LOD) and Limit of Quantitation (LOQ) for each of the analyzed elements using the soil FP calibration method were provided by Bruker (*29*). Those most applicable to this study are shown in Table I. The method was optimized for a SiO_2 matrix and a 120 s analysis time. For quantitation, the Limit of Quantitation was defined as five times the LOD for this specific method.

Table I. Detection Limits for selected elements with the S1 Turbo handheld XRF analyzer using the soil FP calibration method (*29*)

	K	*Ca*	*Ti*	*Mn*	*Fe*	*Cu*	*Zn*	*As*	*Sr*	*Pb*	*Zr*	*Cd*	*Hg*
LOD[a] (ppm)	450	200	90	18	17	3	3	2	3	7	3	12	2
LOQ[b] (ppm)	2250	1000	450	90	85	15	15	10	15	35	15	60	10

[a] LOD, Limit of Detection   [b] LOQ, Limit of Quantitation

Nine edible clay powder supplements (Table II) were analyzed with a Bruker S_1 TURBO handheld XRF instrument as described for the soil standards. All of the samples were from different suppliers. Each of the nine clay products were measured out to five equal samples (0.80 g) and analyzed for 120 s. The five samples were taken at random from the bulk sample of a given product. All but one of the clays purchased were in powdered form. Clay sample #1 came in capsule form and the powdered clay was removed from the capsule for the analysis. The daily exposure for an element was determined from the measured concentration and the daily serving given by the supplier. This value was then compared to the permitted daily exposure (PDE) level as specified by the U.S. Pharmacopeia Forum (*17*).

Results and Discussion

Nine edible clay products (Table II) were analyzed with the handheld XRF analyzer using the soil FP calibration. All of the clay products were advertised as edible products and all but sample #5 stated a specific serving size in their instructions. A conservative serving size of 2.00 g was assumed for calculations for #5 since this was below most of the serving sizes suggested by the other suppliers.

Table II. Clay Sample Information

ID	Product Name	Source	Type of Clay	Serving size (g)	Origin
1	Pascalite	Pascalite, Inc.	Calcium Bentonite	2.15	WY, USA
2	Terramin	California Earth Minerals	Calcium Montmorillonite	3.36	CA, USA
3	Redmond Clay	Redmond Trading Company, LC	Bentonite Clay (Montmorillonite Family)	2.74	UT, USA
4	Edible Earth: Formula No. 1	LL Magnetic Clay, Inc.	Calcium Bentonite; Calcium Montmorillonite	4.50	North America
5	Pure Cosmetic Clay	Mountain Rose Herbs	Bentonite Clay	2.00[a]	Various locations
6	Tecopia Essential Edible Clay Melange	Green Clays	Red & Green Calcium Bentonite: Green Sodium Bentonite	10.67	AZ, USA
7	Bentonite Powder	Now Foods	Sodium Bentonite	1.37	North America
8	Red Desert Clay	Abundance Enterprises, Inc.	Calcium Montmorillonite	6.61	AZ, USA
9	Living Clay	The Living Clay Co.	Calcium Bentonite	5.16	USA

[a] Since no specific serving size was provided by the packaging, company website, or personnel, a value of 2.00 g was used for calculations.

NIST standards were used to investigate the accuracy of the soil FP calibration method using the Bruker S_1 TURBO handheld XRF spectrometer. The values (average ± standard deviation) obtained with the XRF are reported in ppm and include the percent relative standard deviation (in parentheses) as shown in Tables III-V. The averages were compared to the certified or reference values for three soil SRMs. The percent error as compared to the certified or reference values are also given (30–32). SRM 2709 is the San Joaquin soil standard. SRM 2586 is the Trace Elements in Soil standard containing Lead from Paint (Nominal 500 mg lead/kg). SRM 2587 is the Trace Elements in Soil standard containing Lead from Paint (Nominal 3000 mg lead /kg) standard. The results for the concentrations of calcium, iron, and titanium as determined by XRF are compared to the values reported for the SRMs as shown in Table III. The percent errors for calcium were the highest while those for iron and titanium were less than 13%.

Table III. Concentrations of calcium, iron, and titanium from analysis with the handheld XRF as compared to the NIST reported values[a]

	Ca (ppm)	% error[b]	Fe (ppm)	% error	Ti (ppm)	% error
2709 XRF[c]	16900 ± 310 (1.8)[d]	10.5	32520 ± 370 (1.1)	7.1	3000 ± 100 (3.3)	12.4
2709 STD[e]	18900 ± 500		35000 ± 1100		3420 ± 240	
2586 XRF	17680 ± 440 (2.5)	20.3	52660 ± 460 (0.87)	1.9	5790 ± 70 (1.2)	4.3
2586 STD	22180 ± 540		51660 ± 890		6050 ± 68	
2587 XRF	8030 ± 170 (2.1)	13.4	28980 ± 550 (1.9)	3.0	3950 ± 77 (1.9)	0.8
2587 STD	9270 ± 200		28130 ± 30		3920 ± 650	

[a] All values for STD 2709 are certified values and the values for STD 2587 and 2586 are reference values. [b] Percent error between measured XRF value and NIST reported value. [c] XRF, values (average ± standard deviation) obtained from analysis with handheld XRF. [d] Percent relative standard deviation. [e] STD, NIST values (30–32)

The results for the concentrations of lead, manganese, and zinc as determined by XRF were compared to those reported for the SRMs as shown in Table IV. Two of these standard materials (SRM 2586 and 2587) are soil samples where lead paint has been added. The percent errors for results from the XRF as compared to these three soil standards were less than 13%. The percent error for lead in 2709 was not calculated since the value obtained from the XRF for 2709 (10.9 ± 9.9 ppm) was below the LOQ of 35 ppm (Table I).

The concentrations of strontium, copper, and arsenic as determined by XRF were compared to those reported for the SRMs as shown in Table V. The percent errors for the strontium analyses are all below 7%. For the copper analysis, the percent errors ranged from 3.2% for 2586 and 21.9% for 2709. No comparison was made for the copper in 2709 since the standard value given for SRM 2709 was an informational value and not a certified or reference value. The percent errors for the arsenic analysis were 33-40%. The average values obtained from the XRF are very close to the LOQ.

None of the SRMs had reported cadmium levels that were above the LOD of 12 ppm, and no cadmium was detected in these samples when analyzed by XRF. In addition, none of the SRMs had reported mercury levels above the LOD of 2 ppm, and no mercury was reported for the samples when analyzed by XRF.

Table IV. Concentrations of lead, manganese, and zinc from analysis with the handheld XRF as compared to the NIST reported values[a]

	Pb (ppm)	% error[b]	Mn (ppm)	% error	Zn (ppm)	% error
2709 XRF[c]	10.9 ± 9.9 (90)[d]	ND[f]	470 ± 25 (5.3)	12.5	110 ± 6 (5)	3.8
2709 STD[e]	18.9 ± 0.5		538 ± 17		106 ± 3	
2586 XRF	451 ± 22 (4.9)	4.3	913 ± 28 (3.1)	8.6	378 ± 24 (6.3)	7.3
2586 STD	432 ± 17		1000 ± 18		352 ± 6	
2587 XRF	3570 ± 275 (7.7)	1.8	597 ± 25 (4.2)	8.2	419 ± 74 (18)	2.4
2587 STD	3242 ± 57		651 ± 23		335 ± 8	

[a] All values for STD 2709 and Pb values for 2586 and 2587 are certified values. All others are reference values. [b] Percent error between measured XRF value and NIST reported value. [c] XRF, values (average ± standard deviation) obtained from analysis with handheld XRF. [d] Percent relative standard deviation. [e] STD, NIST values (*30–32*). [f] ND, not determined since the XRF value was less than the LOQ.

The results from the XRF analysis for the concentrations of various elements in the nine clay samples are shown in Figures 1-3. The comparison of the potassium, calcium and iron levels are shown in Figure 1. Figure 2 shows the results from the analysis for titanium, manganese and strontium. Figure 3 shows the results from the analysis for copper, zinc and zirconium.

The clay samples were analyzed for lead and results are shown in Table VI. All of the samples contained levels of lead that were above the LOD of 7 ppm. None of the samples had values that were above the LOQ. The daily exposure was not calculated for these samples that were above the LOD but below the LOQ. The permitted daily exposure (PDE) for lead in supplements is 10 µg/day (USP) (*17*).

The clay samples were analyzed for arsenic and the concentration and daily exposure amounts are shown in Table VII. All of the samples contained levels of arsenic that were above the LOD of 2 ppm. Eight of the samples contained amounts of arsenic that were also above the LOQ. Using these values there is the possibility that samples 2-4 and 6-9 would all result in a daily exposure to inorganic arsenic that is above the permitted daily exposure (PDE) of 15 µg/day (*17*). The form of arsenic was not determined from this analysis. It would be of interest to look at the samples that had higher concentrations of lead and arsenic by a different analytical method in order to determine if these clay supplements would in fact result in a daily exposure which exceeds the FDA permitted daily exposure limits. The XRF analysis using the soil FP calibration did not detect cadmium or mercury in the clay samples.

Table V. Concentrations of strontium, copper, and arsenic from analysis with the handheld XRF as compared to the NIST reported values[a]

	Sr (ppm)	% error[b]	Cu (ppm)	% error	As (ppm)	% error
2709 XRF[c]	247 ± 6 (2.4)[d]	6.9	42.2 ± 6 (14)	21.9	11.8 ± 6.6 (56)	33.3
2709 STD[e]	231 ± 2		34.6 ± 0.7		17.7 ± 0.8	
2586 XRF	82.2 ± 3.3 (4.9)	2.3	78.4 ± 8.0 (10)	3.2	12.2 ± 3.3 (27)	40.2
2586 STD	84.1 ± 8		81[a]		8.7± 1.5	
2587 XRF	134 ± 8 (6)	6.7	12 ± 7 (58)		18.5 ± 9.7 (52)	35
2587 STD	126 ±19		[f]		13.7 ± 2.3	

[a] All values for STD 2709 are certified values. All others are reference values.   [b] Percent error between measured XRF value and NIST reported value.   [c] XRF, values (average ± standard deviation) obtained from analysis with handheld XRF.   [d] Percent relative standard deviation.   [e] STD, NIST values (30–32)   [f] Only informational mass values were given for this element.

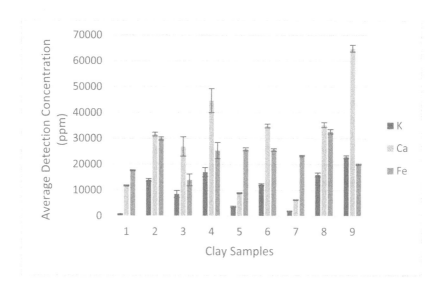

Figure 1. The concentrations (average ± standard deviation) of potassium, calcium and iron in the clay samples as determined by XRF analysis.

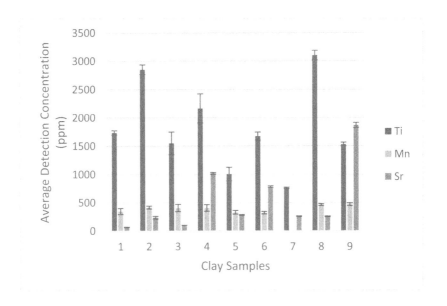

Figure 2. The concentrations (average ± standard deviation) of titanium, manganese, and strontium in the clay samples as determined by XRF analysis.

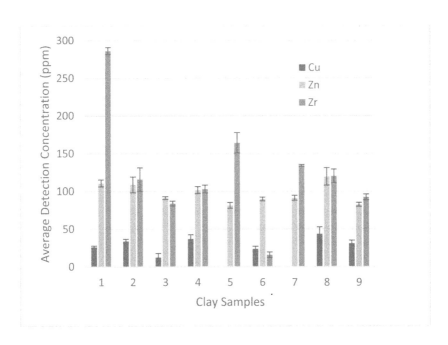

Figure 3. The concentrations (average ± standard deviation) of copper, zinc, and zirconium in the clay samples as determined by XRF analysis.

Table VI. Concentration of lead in the clay supplements

ID	Lead (ppm)[a]	Above LOQ[b]	Daily Exposure (µg/day)
1	26.6 ± 1.8 (6.8)[c]	No	ND[d]
2	20.2 ± 4.3 (21)	No	ND
3	16.2 ± 2.7 (17)	No	ND
4	15.4 ± 2.3 (15)	No	ND
5	30.6 ± 5.0 (16)	No	ND
6	19.2 ± 2.4 (13)	No	ND
7	32.2 ± 5.8 (18)	No	ND
8	19.8 ± 4.8 (24)	No	ND
9	14.8 ± 2.0 (14)	No	ND

[a] Average ± Standard deviation. [b] LOQ, Limit of Quantitation, 35 ppm [c] Percent relative standard deviation [d] ND, not determined

Table VII. Concentration of arsenic in the clay supplements

ID	Arsenic (ppm)[a]	Above LOQ[b]	Daily Exposure (µg/day)
1	3.0 ± 1.4 (46)[c]	No	ND[d]
2	11.4 ± 2.0 (18)	Yes	38.3
3	13.4 ± 2.0 (15)	Yes	36.7
4	17.6 ± 1.7 (10)	Yes	79.2
5	7.2 ± 0.8 (11)	No	ND
6	17.8 ± 2.3 (13)	Yes	190
7	14.2 ± 1.8 (13)	Yes	44.1
8	12.6 ± 2.6 (21)	Yes	19.5
9	23.4 ± 2.0 (8.5)	Yes	83.3

[a] Average ± Standard deviation [b] LOQ, Limit of Quantitation, 10 ppm [c] Percent relative standard deviation [d] ND, not determined

Conclusions

A handheld XRF using a soil FP calibration method was used to analyze three soil NIST SRMs and the results were compared to the values reported for each. The results from the analysis of the soil SRMs with the handheld XRF were within 13% of the reported values for iron, titanium, lead, manganese, zinc, and strontium and 20% for copper and calcium. The percent errors for arsenic were larger. The large percent errors are not unexpected given how close to the LOQ the average values are and the large relative standard deviations for these measurements. No cadmium or mercury was detected in the soil standards and this was expected since the reported values for all samples were below the LODs for these.

The seven clay samples had different compositions as determined by XRF analysis using the soil FP calibration method. The metals that would be of most concern would be arsenic, lead, cadmium and mercury. Neither cadmium nor mercury was detected in any of the clay samples. Lead was found in all of the clay samples but none had levels that were above the LOQ. All of the clay samples had detectable levels of arsenic. It is possible that ingesting eight of these samples in the expected daily serving could result in levels of arsenic that exceed the permitted daily exposure limits for inorganic arsenic (*17*). This is a simple, rapid, non-destructive, and low-cost method for analyzing these clay supplements and gives initial results which can be further investigated with other instrumental methods that require more sample preparation time and more costly analysis. It would be important to further analyze the samples that had higher concentrations of the lead and arsenic by another analytical method to verify these results and to also determine the form of arsenic. Finally it would be of interest to determine the bioavailability of these metals after ingestion.

Acknowledgments

Funding for this project was made possible by the University of Detroit Mercy's McNichols Faculty Assembly Internal Research Fund and the Michigan-Ohio University Transportation Center Grant.

References

1. A Union of the Finest Internal Clay Supplements. Tecopia Essential Edible Clay Mélange. www.greenclays.com/clay-supplements.php (accessed April 18, 2012).
2. Using Bentonite, Illite, and Montmorrilonite Healing Clays Internally, 2012. Entons' Earth. www.eytonsearth.org/drinking-clay-internal-use.php (accessed April 18, 2012).
3. Katsoufis, C. P.; Kertis, M.; McCullough, J.; Pereira, T.; Seeherunvong, W.; Chandar, J.; Zilleruelo, G.; Abitbol, C. Pica: an important and unrecognized problem in pediatric dialysis patients. *J. Renal Nutr.* **2012**, *6*, 567–71.
4. Young, S. L.; Sherman, P. W.; Lucks, J. B.; Pelto, G. H. Why on earth?: Evaluating hypotheses about the physiological functions of human geophagy. *Q. Rev. Biol.* **2011**, *2*, 97–120.
5. Federal Food, Drug, and Cosmetic Act. Public Law 75-717; 1938.
6. Dietary Supplement Health and Education Act of 1994. Public Law 103-417; 1994.
7. *Dietary Supplements: FDA Should Take Further Actions to Improve Oversight and Consumer Understanding*; GAO-09-250; United States Government Accountability Office Report to Congressional Requesters, Government Accountability Office: Washington, DC, 2009.

8. Saper, R. B.; Stefanos, N. K.; Paquin, J.; Burns, M. J.; Eisenberg, D. M.; Davis, R. B.; Philips, R. S. Heavy Metal Content of Ayurvedic Herbal Medicine Products. *J. Am. Med. Assoc.* **2004**, *292*, 2868–2873.

9. Saper, R. B.; Philips, R. S.; Seghal, A.; Khouri, N.; Davis, R. B.; Paquin, J.; Thuppil, V.; Kales, S. N. Lead, Mercury, and Arsenic in US- and Indian-Manufactured Ayurvedic Medicines Sold via the Internet. *J. Am. Med. Assoc.* **2008**, *300*, 915–923.

10. Woolf, A. D.; Hussain, J.; McCullough, L.; Petranovic, M.; Chomchai, C. Infantile lead poisoning from an Asian tongue powder: A case report & subsequent public health inquiry. *Clin. Toxicol.* **2008**, *46*, 841–846.

11. Dolan, S. P; Nortrup, D. A.; Bolger, M.; Capar, S. G. Analysis of Dietary Supplements for Arsenic, Cadmium, Mercury, and Lead Using Inductively Coupled Mass Spectrometry. *J. Agric. Food Chem.* **2003**, *51*, 1307–1312.

12. Levine, K. E.; Levine, M. A.; Weber, F. X.; Hu, Y.; Perlmutter, J.; Grohse, P. M. Determination of Mercury in an Assortment of Dietary Supplements Using an Inexpensive Combustion Atomic Absorption Spectrometry Technique. *J. Autom. Methods Manage. Chem.* **2005**, *4*, 211–216.

13. Amster, E.; Tiwary, A.; Schenker, M. B. Case Report: Potential Arsenic Toxicosis Secondary to Herbal Kelp Supplement. *Environ. Health Perspect.* **2007**, *115*, 606–608.

14. Garshott, D. M.; Macdonald, E. A.; Murray, M. N.; Benvenuto, M. A.; Roberts-Kirchhoff, E. Elemental Analysis of a Variety of Dried, Powdered, Kelp Food Supplements for the Presence of Heavy Metals via Energy-Dispersive X-Ray Fluorescence Spectrometry. In *It's All In the Water: Studies of Materials and Conditions in Fresh and Salt Water Bodies*; Benvenuto, M. A., Roberts-Kirchhoff, E. S., Murray, M. N., Garshott, D. M., Eds.; ACS Symposium Series; American Chemical Society: Washington, D.C., 2011; Vol. 1086, pp 123−133.

15. Cohen, P. A.; Ernst, E. Safety of Herbal Supplements: A Guide for Cardiologists. *Cardiovasc. Ther.* **2010**, *28*, 246.

16. Buettner, C.; Mukamal, K. J.; Gardiner, P.; Davis, R. B.; Phillips, R. S.; Mittleman, M. A. Herbal supplement use and blood lead levels of United States adults. *J. General Intern. Med.* **2009**, *24*, 1175–1182.

17. The United States Pharmacopeial Convention. Pharmacopeial Forum. Elemental Contaminants in Dietary Supplements (2232). *Fed. Regist.* **2010**, *36*, 1–9.

18. Heyman, A.; Pfeiffer, J. B.; Willet, R. W.; Taylor, H. M. Peripheral neuropathy caused by arsenic intoxication; a study of 41 cases with observations on the effects of BAL (2,3 dimercapto-propanol). *N. Engl. J. Med.* **1956**, *254*, 401–409.

19. National Institutes of Environmental Health Sciences. Environmental Health Topics: Lead. http://www.niehs.nih.gov/health/topics/agents/lead/index.cfm (accessed September 17, 2013).

20. National Institutes of Environmental Health Sciences. Environmental Health Topics: Mercury. http:// www.niehs.nih.gov/health/topics/agents/mercury/index.cfm (accessed September 17, 2013).

21. Hou, X. D.; He, Y. H.; Jones, B. T. Recent advances in portable X-ray fluorescence spectrometry. *Appl. Spectrosc. Rev.* **2004**, *39*, 1–25.

22. Palmer, P.; Jacobs, R.; Baker, P. E.; Ferguson, K.; Webber, S. Use of Field-Portable XRF Analyzers for Rapid Screening of Toxic Elements in FDA-Regulated Products. *J. Agric. Food Chem.* **2009**, *57*, 2605–2613.

23. Block, C. N.; Shibata, T.; Solo-Gabriele, H. M.; Townsend, T. G. Use of handheld X-ray fluorescence spectrometry for identification of arsenic in treated wood. *Environ. Pollut.* **2007**, *148*, 627–633.

24. Carr, R.; Zhang, C.; Moles, N.; Harder, M. Identification and mapping of heavy metal pollution in soils of a sports ground in Galway City Ireland, using a portable XRF analyzer and GIS. *Environ. Geochem. Health* **2008**, *30*, 45–52.

25. Markey, A. M.; Clark, C. S.; Succop, P. A.; Roda, S. Determination of the feasibility of using a portable X-ray fluorescence (XRF) analyzer in the field for measurement of lead content of sieved soil. *J. Environ. Health* **2008**, *70*, 24–29.

26. Taylor, A.; Day, M. P.; Marshall, J.; Patriarca, M.; White, M. Atomic spectrometry update. Clinical and biological materials, food and beverages. *J. Anal. At. Spectrom.* **2012**, *27*, 537–576.

27. Anderson, D. L. Analysis of beverages for Hg, As, Pb, and Cd with a field portable X-ray fluorescence analyzer. *J. AOAC Int.* **2010**, *93*, 683–693.

28. Anderson, D. L. Screening of foods and related products for toxic elements with a portable X-ray tube analyzer. *J. Radioanal. Nucl. Chem.* **2009**, *282*, 415–418.

29. *SI Turbo^sdr Calibrations*; Bruker Elemental: Kennewick, WA, November 2010.

30. Wise, S. A.; Watters, R. L. *Certificate of Analysis SRM®2586 Trace Elements in Soil Containing Lead from Paint (Nominal 500 mg/kg Lead)*; National Institute of Standards & Technology: Gaithersburg, MD, 2008.

31. Wise, S. A.; Watters, R. L. *Certificate of Analysis SRM®2587 Trace Elements in Soil Containing Lead from Paint (Nominal 3000 mg/kg Lead)*; National Institute of Standards & Technology: Gaithersburg, MD, 2008.

32. Wise, S. A.; Watters, R. L. *Certificate of Analysis SRM®2709a San Joaquin Soil*; National Institute of Standards & Technology: Gaithersburg, MD, 2008.

Editors' Biographies

Mark A. Benvenuto

Mark Benvenuto is a Professor of Chemistry at the University of Detroit Mercy, in the Department of Chemistry & Biochemistry. His research thrusts span a wide array of subjects, but include the use of energy dispersive X-ray fluorescence spectroscopy to determine trace elemental compositions of: aquatic and land-based plant matter, food and dietary supplements, and medieval and ancient artifacts.

Benvenuto received a B.S. in chemistry from the Virginia Military Institute, and after several years in the Army, a PhD. in inorganic chemistry from the University of Virginia. After a post-doctoral fellowship at the Pennsylvania State University, he joined the faculty at the University of Detroit Mercy in 1993

Satinder Ahuja

Satinder Ahuja (Ph. D. from the University of Sciences in Philadelphia) worked at Novartis Corporation for over 25 years in various leadership positions and simultaneously served as adjunct professor for several universities. For the last decade, he has been helping solve water contamination problems worldwide. As a founder of Ahuja Academy of water Quality at UNC Wilmington, he encourages research on various water quality issues. His latest books include *Monitoring Water Quality* (Elsevier 2013), *Novel Solutions to Water Pollution* (ACS 2013), *Comprehensive Water Quality and Purification* (Elsevier 2013), *Handbook of Water Purity and Quality* (Elsevier 2009), and *Arsenic Contamination of Water: Mechanism, Analysis, and Remediation* (Wiley 2008).

Timothy V. Duncan

Timothy Duncan is a research scientist at the Center for Food Safety and Applied Nutrition, part of the U.S. Food and Drug Administration. His primary research focus is safety of nanomaterials utilized in foods and food contact materials. Dr. Duncan received his B.S. in chemistry from Haverford College and his Ph.D. in physical/inorganic chemistry from the University of Pennsylvania, where he also completed a post-doctoral fellowship. He has been with FDA since 2009.

Gregory O. Noonan

Gregory Noonan is a Research Chemist in the Office of Regulatory Science in the Center of Food Safety and Applied Nutrition at the US Food and Drug Administration. He works on the development of methods for the determination of food additives and contaminants in foods and food contact materials. Gregory Noonan received his B.S. in chemistry from SUNY Albany and his M.S and Ph.D. in chemistry from Michigan State University. He also worked as a postdoctoral fellow in the Department of Civil Environmental Engineering at the Massachusetts Institute of Technology.

Elizabeth S. Roberts-Kirchhoff

Elizabeth S. Roberts-Kirchhoff is Associate Professor of Chemistry and Biochemistry at the University of Detroit Mercy. Her research interests include the mechanism of action of cytochrome P450 enzymes including their role in the metabolism of drugs and natural products and the investigation of heavy metals in health supplements including kelp, clay, and protein powders.

Roberts-Kirchhoff received a B.S. in Chemistry from Texas A & M University and a Ph.D. in Biological Chemistry from the University of Michigan. After postdoctoral research at Wayne State University and The University of Michigan, she joined the faculty at the University of Detroit Mercy in 1997.

Indexes

Author Index

Ahuja, S., 1
Benvenuto, M., 1, 87, 99
Bott, J., 51
Crawford, E., 71
Croce, T., 41
Crone, C., 71
Diefenbach, A., 87
Duncan, T., 1
Fahey, M., 87
Franz, R., 51
Garshott, D., 87
Horner, J., 71
Kleinschmidt, L., 5

LaBond, J., 99
MacDonald, E., 87
Maurice, S., 87
Musselman, B., 71
Noonan, G., 1
Roberts-Kirchhoff, E., 1, 99
Sanday, T., 87
Seiber, J., 5
Störmer, A., 51
Stroeters, N., 99
Tongesayi, S., 15
Tongesayi, T., 15

Subject Index

C

CID. *See* Collision induced dissociation (CID)
Collision induced dissociation (CID), 71

D

DART. *See* Direct analysis in real time (DART)
Direct analysis in real time (DART), 71

E

Essential elements and human health
  manganese, 23
  zinc, 24
Extracted ion chromatograms (XICs), 82*f*

F

Food distribution system, 1
Food packaging, 71
Food production, distribution and waste, 3
Food safety and security
  overview, 1
  technologies, 2
Food web, 1

G

Global contamination of food
  chemical exposure, 18
  chronic diseases, 17
  environmental pollutants, 15
  food imports, 18
  global citizenry, 17
  globalized food market and need for diverse foods, 17
  introduction, 16
  toxic and priority pollutant metal(loid)s, 19

H

Heavy metal(loid)s and human health
  arsenic, 19
  cadmium, 20
  lead, 21
    main exposure sources, 22
    pre-, peri-, and postnatal exposure, 22
Heavy metal(loid)s in food
  agricultural practices and food contamination
    anthropogenic activities, 29
    Chinese electronic media, 29
    composition of sewage, 28
    Czech Republic, 30
    extent of food contamination, 29
    irrigation of crops, use of sewage and industrial effluents, 28
    scientific literature, 28
    state secret, 30
    Thailand, contaminated rice, 30
  electronic waste (e-waste), 26
    China, 27
    global production, 27
    recycling impact on soil, 27
  extensive use of agrochemicals, 25
  hazardous waste, 25
  rice and rice food products, 24

M

Migration potential of nano silver particles
  10 % ethanol, 60*f*
  AF4/MALLS measurements, 65
  Ag-NPs dispersion, stability testing, 61
  approaches, 66
  Asymmetric Flow Field-Flow Fractionation (AF4), 55
  conclusion, 67
  food contact polyolefins, 51
  Inductively Coupled Plasma Mass Spectrometry (ICP-MS), 54
  introduction, 52
  materials, 53
  migration experiments, 54
  migration modelling, 56, 62
  migration of silver from LDPE films, 59
  migration of silver into 3 % acetic acid, 60*f*

modeled migration of spherical carbon NPs from LDPE, 64*f*

Multi Angle Laser Light Scattering Spectrometry (MALLS), 56

nanomaterial characterization in polymer by TEM, 57

physico-chemical specifications, 63*t*

recovery experiments and detection limit of silver by ICP-MS, 58

silver content of LDPE films, determination by acid digestion, 54, 58

Transmission Electron Microscopy (TEM), 53

unambigous measurement, difficulties, 67

Dietary Supplement Health and Education Act of 1994 (DSHEA), 100

dietary supplement manufacture, 100

experimental methods, 101

Federal Food, Drug, and Cosmetic Act, 100

handheld XRF as compared to NIST reported values

concentrations of calcium, iron, and titanium, 104*t*

concentrations of lead, manganese, and zinc, 105*t*

concentrations of strontium, copper, and arsenic, 106*t*

introduction, 99

results and discussion, 102

soil FP calibration method, 103

N

Nanotechnology in food ingredients

conclusion, 48

food ingredients in United States, regulation, 42

introduction, 41

regulation, FDA's approach, 43

adaptive approach, 44

altered properties, 44

food ingredient authorizations, 46

food ingredients, characterization, 46

food ingredients, safety standard, 44

guidance, 43

impact of manufacturing changes on regulatory status, 47

National Nanotechnology Initiative (NNI) Program, 44

product-specific draft guidance, 45

regulatory status, 46

toxicology considerations, 47

Nine edible clay supplements

chemical composition, 100

clay sample information, 103*t*

clay samples as determined by XRF analysis

concentrations of copper, zinc, and zirconium, 107*f*

concentrations of potassium, calcium and iron, 106*f*

concentrations of titanium, manganese, and strontium, 107*f*

concentration of arsenic, 108*t*

concentration of lead, 108*t*

conclusions, 108

detection limits for selected elements, 102*t*

P

Patent medicines

beneficial elements, 91

certified AAS standard solutions, 89

common elements, 93

conclusions, 95

deleterious elements, 91

Dr. Sawen's Magic Nervine Pills box, 89*f*

experimental, 88

Henry Ford Museum, 87

introduction, 87

other elements, 94

results and discussion, 89

samples, elements, relative amounts at 20keV, 90*f*

samples containing lead, mercury, and / or arsenic, 92*f*

Phthalate monitoring in food stuffs, 71

airline plastic fork, XICs and related MS/MS spectra sampling, 82*f*

CID and HCD fragmentation, 77*f*

CID MS/MS spectral fingerprints, 78*f*

conclusions, 83

experimental set-up, 73

DART-SVP and Velos Pro MS, 74

DART-SVP ion source, 76*f*

higher-energy collisional dissociation (HCD) fragmentation, 73

low energy CID fragmentation, 73

related EU legislation, 75*t*

introduction, 72

isomeric phthalates, predicted fragmentation pathways, 79*f*

method development, 76

method evaluation, 79
MS method development, 77
screened consumer products and detected phthalates, 81*t*
screening consumer products and food packaging, 80
Pragmatic approaches to food contamination
enforceable safe limits, 32
European Food Safety Authority (EFSA), 32
extent of pollution, 33
food testing and safe limits, 31
prevention may be the key
comprehensive testing, 33
crafting and policing international laws, 34
curtailing pollution, 33
unethical agricultural practices, 34
water shortages, 34
tolerable weekly intake (TWI), 33

R

Regulation of food ingredients in the United States
food additives and color additives, 42
food contact notifications, 43
GRAS substances, 43

S

Sustainability in foods and food production
conclusion, 11

consumers' view, 10
corporate involvement, 9
current interest, 5
recent research, examples and challenges, 10
safety aspects, 8
U.S. food and beverage industry, 6
view from processors, 7
view from producers, 7
world food production, 5

T

Total Diet Study Program, 32
Toxic Elements Program, 32

U

U.S. food and beverage industry, foods and food production, sustainability, 5

W

World food production, foods and food production, sustainability, 5

X

XICs. *See* Extracted ion chromatograms (XICs)